Craft Beverages and Tourism, Volume 2

Susan L. Slocum • Carol Kline • Christina T. Cavaliere
Editors

Craft Beverages and Tourism, Volume 2

Environmental, Societal, and Marketing Implications

Editors
Susan L. Slocum
George Mason University
Manassas, Virginia, USA

Carol Kline
Appalachian State University
Raleigh, North Carolina, USA

Christina T. Cavaliere
Stockton University
Manahawkin, New Jersey, USA

ISBN 978-3-319-57188-1   ISBN 978-3-319-57189-8 (Ebook)
DOI 10.1007/978-3-319-57189-8

Library of Congress Control Number: 2017945308

© The Editor(s) (if applicable) and The Author(s) 2018
This work is subject to copyright. All rights are solely and exclusively licensed by the Publisher, whether the whole or part of the material is concerned, specifically the rights of translation, reprinting, reuse of illustrations, recitation, broadcasting, reproduction on microfilms or in any other physical way, and transmission or information storage and retrieval, electronic adaptation, computer software, or by similar or dissimilar methodology now known or hereafter developed.
The use of general descriptive names, registered names, trademarks, service marks, etc. in this publication does not imply, even in the absence of a specific statement, that such names are exempt from the relevant protective laws and regulations and therefore free for general use.
The publisher, the authors and the editors are safe to assume that the advice and information in this book are believed to be true and accurate at the date of publication. Neither the publisher nor the authors or the editors give a warranty, express or implied, with respect to the material contained herein or for any errors or omissions that may have been made. The publisher remains neutral with regard to jurisdictional claims in published maps and institutional affiliations.

Cover illustration: © Monty Rakusen / Getty Images

Printed on acid-free paper

This Palgrave Macmillan imprint is published by Springer Nature
The registered company is Springer International Publishing AG
The registered company address is: Gewerbestrasse 11, 6330 Cham, Switzerland

ACKNOWLEDGMENTS

The editors would like to thank the following for their support in this project:
Ken Backman, Professor, Clemson University
Karla Boluk, Assistant Professor, University of Waterloo
Rachel Chen, Professor, University of Tennessee
Kynda Curtis, Professor, Utah State University
Gerrie Du Rand, Senior Lecturer, University of Pretoria
Craig Esherick, Associate Professor, George Mason University
Sally Everett, Deputy Dean, Quality and Student Experience, Anglia Ruskin University
John Hull, Associate Professor, Thompson Rivers University
Deb Kerstetter, Professor, The Pennsylvania State University
Geoff Lacher, Senior Economist, Oxford Economics
Dominic Lapoint, Professor, Université du Québec à Montréal
Shawn Lee, Associate Professor, George Mason University
Leah Mathews, Professor, University of North Carolina, Asheville
Liz Sharples, Lecturer, University of Portsmouth
Dawn Thilmany, Professor, Colorado State University
Dallen Timothy, Professor, Arizona State University
Mark Wickham, Senior Lecturer, University of Tasmania
Brenda Wiggins, Associate Professor, George Mason University
Peter Wiltshier, Senior Lecturer, University of Derby
Aaron Yonkholmes, Assistant Professor, Macau University of Science and Technology

Contents

1 Introduction 1
Susan L. Slocum

2 Brewing Green: Sustainability in the Craft Beer
Movement 9
Ellis Jones

3 Craft Beer Enthusiasts' Support for Neolocalism
and Environmental Causes 27
David Graefe, Andrew Mowen, and Alan Graefe

4 Pure Michigan Beer? Tourism, Craft Breweries,
and Sustainability 49
Michael J. Lorr

5 Representing Rurality: Cider Mills and Agritourism 65
Wynne Wright and Weston M. Eaton

6 Developing Social Capital in Craft Beer Tourism Markets 83
Susan L. Slocum

7	New Jersey Craft Distilleries: Sense of Place and Sustainability Christina T. Cavaliere and Donna Albano	101
8	Drink Tourism: A Profile of the Intoxicated Traveler Kynda R. Curtis, Ryan Bosworth, and Susan L. Slocum	119
9	Craft Brewing Festivals Zachary M. Cook	141
10	(Micro)Movements and Microbrew: On Craft Beer, Tourism Trails, and Material Transformations in Three Urban Industrial Sites Colleen C. Myles and Jessica McCallum Breen	159
11	Brewing a Beer Industry in Asheville, North Carolina Scott D. Hayward and David Battle	171
12	An Exploration of the Motivations Driving New Business Start-up in the United States Craft Brewing Industry Erol Sozen and Martin O'Neill	195
13	Conclusion Susan L. Slocum, Christina T. Cavaliere, and Carol Kline	213
Index		225

Notes on Contributors

Donna Albano is Associate Professor in the Hospitality and Tourism Management Studies (HTMS) Program at Stockton University's School of Business. Donna holds Doctorate from Rowan University in Glassboro, NJ and was the recipient of the Larry Marcus Award for Excellence in Educational Leadership, Doctoral Studies in 2007 and most recently awarded the 2016 Spirit or Hospitality Award given to an outstanding individual who has made a long-lasting and significant contribution to the Atlantic City region's hospitality and travel industry.

David Battle graduated from Cornell University with a BA in Economics and has an MBA from Appalachian State University Walker College of Business. Dave was a general contractor in the residential construction industry through 2011 and witnessed the growth of the craft brewing industry in Asheville firsthand. He is employed in Global Technology and Operations at Bank of America, Charlotte.

Ryan Bosworth is an Associate Professor in the Department of Applied Economics at Utah State University. He received his Ph.D. from the University of Oregon in 2006. He has also received MS and BS degrees in Economics from Utah State University. He specializes in applied micro-econometrics.

Jessica Breen is a Ph.D. Candidate in the Geography Department of the University of Kentucky. She is a cartographer and urban geographer whose research explores the role of art and creativity in urban creative placemaking efforts.

Christina T. Cavaliere is an environmental social scientist and international sustainable development specialist focused on linking tourism and bio-cultural conservation. She serves as Assistant Professor of Hospitality and Tourism Management and Sustainability at Stockton University. Her research interests include tourism and climate change, local economies, sustainable agriculture and ecogastronomy, permaculture, agritourism, and community redevelopment.

Zachary M. Cook holds a bachelor's degree in history from Lebanon Valley College in 2006 and his master's degree from Millersville University of Pennsylvania. As an avid lover of local history, Zachary's research includes work on early regional automobile manufacturing, veterans of the Civil War as well as historical changes in brewing techniques. Zachary is a doctoral student of Pennsylvania State University and his research involves beer and the history of brewing beer in Pennsylvania, especially as it relates to Pennsylvania-Germans and modern-day beer festivals.

Kynda Curtis is a Professor and Extension Economist in the Department of Applied Economics at Utah State University. Her research interests include international agriculture/food marketing, consumer economics, and industrial organization. In her extension appointment, she works with agricultural producers to assist them in developing new markets for their products and assessing the feasibility of new food and agricultural products and value-added processes.

Weston M. Eaton is a Postdoctoral Research Associate in the Agricultural Economics, Sociology and Education (AESE) Department at the Pennsylvania State University. His research bridges science studies and social movements studies, and draws from environmental sociology and cultural sociology, to examine how people and communities construct meaning and take action on energy, environmental and agricultural issues. Born in Michigan, Weston is also a hobbyist beer and wine brewer and has won multiple awards for oak aged hard cider made from Michigan apples.

Alan Graefe teaches courses and conducts research about outdoor recreation behavior and management at the Pennsylvania State University. He was one of the developers of the Visitor Impact Management (VIM) framework for addressing visitor capacity issues and was a co-founder of the Northeastern Recreation Research (NERR) Symposium. His principal research interests revolve around the application of social science to various aspects of recreation resources planning and management.

David Graefe is an Assistant Professor of Natural Resources and Recreation Management at Marshall University. In addition to being a craft beer enthusiast, Graefe specializes in outdoor recreation theory, park and protected area management, and the human dimensions of natural resources.

Scott D. Hayward is an Assistant Professor in the Department of Management at Elon University. His research focuses on the micro-mechanisms of strategic management, with a secondary interest in the intersection of economic geography and innovation. He received his MBA and Ph.D. from Emory University in Atlanta, GA.

Ellis Jones is an Assistant Professor of Sociology at the College of the Holy Cross in Worcester, MA. His research is primarily focused on understanding the relationship between ethical consumers and corporate social responsibility. Recently, he has become more specifically interested in analyzing the social and environmental impacts of the craft beer movement. He is the author of *The Better World Shopping Guide* (5th edition) and co-author of *The Better World Handbook* (2nd edition).

Carol Kline is an Associate Professor of Hospitality and Tourism Management in the Department of Management at Appalachian State University. Her research interests focus broadly on tourism planning and development and tourism sustainability, but cover a range of topics such as foodie segmentation, craft beverages, agritourism, wildlife-based tourism, animal ethics in tourism, tourism entrepreneurship, niche tourism markets, and tourism impacts to communities.

Michael J. Lorr is an Associate Professor of Sociology and the Director of the Community Leadership Program at Aquinas College in Grand Rapids, MI. His work on urban, environmental and cultural sociology has recently appeared in the *Journal of Contemporary Ethnography, Nature & Culture, Humanity & Society* and the *Journal of Youth and Adolescence*.

Andrew Mowen is a Professor in the Recreation, Park and Tourism Management Department at Pennsylvania State University. Mowen's work focuses on the contribution of parks and outdoor recreation in shaping public health. He also specializes in understanding the potential and pitfalls of public-private partnerships.

Colleen C. Myles is an Assistant Professor of Geography at Texas State University in San Marcos, TX. She is a rural geographer and political ecologist with specialties in land and environmental management, (ex)urbanization, (rural) sustainability and tourism, wine, beer, and cider geographies (aka "fermented landscapes"), and agriculture (urban, peri-urban, and sustainable).

Martin O'Neill is the Horst Schulze Endowed Professor of Hospitality Management and Head of Auburn University's Hospitality Management Brewing Science Programs. O'Neill has taught a variety of operational and management courses across three continents including Australia, Europe, and the United States. His research interests include services marketing, management, and hospitality operations.

Susan L. Slocum is an Assistant Professor at George Mason University and specializes in sustainable economic development through tourism and policy implementation at the regional and national level. Working with communities to enhance backward linkages between tourism and traditional industries, Slocum has worked with rural communities in the United States, the United Kingdom, and with indigenous populations in emerging tourism destinations in Tanzania.

Erol Sozen is a Ph.D. student and a Graduate Assistant in the Department of Nutrition, Dietetics and Hospitality Management at Auburn University. To date, Erol Sozen has published one research paper and presented many kinds of research in conferences at the national and international levels. Erol Sozen conducts research in the area of craft brewing and customer satisfaction and marketing within the hospitality industry.

Wynne Wright is an Associate Professor jointly appointed in the Departments of Community Sustainability and Sociology at Michigan State University. She is particularly drawn to the study of rural places and people. Much of her research, teaching, and engagement work explores the nexus between agrifood systems, rural culture, and gender. Her research examines questions of political and socio-cultural change in agriculture and food systems with sensitivity to their consequences for farm families and rural community well-being.

LIST OF FIGURES

Fig. 3.1	Craft beer commitment, neolocalism support, and environmentalism	41
Fig. 5.1	Uncle John's Cider Mill billboard image	73
Fig. 5.2	Vander Mill Chapman's Blend label	77
Fig. 6.1	Elements of successful social capital building	86
Fig. 7.1	James F. C. Hyde (1825–1898)	106
Fig. 7.2	Great Notch Distiller logo	110
Fig. 8.1	Distribution of respondent participation in drink tourism	127
Fig. 11.1	Brewmaster movement, 1994–2011	180
Fig. 11.2	Triadic model of local identity interactions	182

LIST OF TABLES

Table 2.1	Rank list of sustainable breweries by mentions in online media	12
Table 2.2	List of sustainable breweries ranked by size with position titles noted	13
Table 3.1	Craft beer travel and visitation patterns	35
Table 3.2	Craft beer "social worlds" behaviors	36
Table 3.3	Craft beer commitment/involvement	37
Table 3.4	Support for brewery neolocalism	38
Table 3.5	Pro-environmental behaviors	39
Table 6.1	Brewery information	88
Table 6.2	Survey demographics	89
Table 6.3	Findings in relation to bridging social capital	97
Table 7.1	Framework indicators, themes, and examples	113
Table 8.1	Sample summary statistics and difference in mean test results	123
Table 8.2	Logit model results—demographic variables	129
Table 8.3	Logit model results—Utah visit rational	130
Table 8.4	Logit model results—activities at home	131
Table 8.5	Logit model results—activities while traveling	133
Table 8.6	Logit model results—Utah knowledge, attitudes, and interests	134
Table 9.1	Distance traveled to reach beer festival (rounded to nearest percent), $N = 236$	148
Table 9.2	How Patrons heard about the beer fest (rounded to nearest percent), $N = 197$	148
Table 9.3	Socioeconomic status of patron (rounded to nearest percent), $N = 189$	150
Table 9.4	Opinions on craft beer (rounded to nearest percent), $N = 196$	151

Table 11.1 Asheville craft brewery growth timeline (2009–2012) 179
Table 12.1 Demographic background 204
Table 12.2 Entrepreneurial motivation variables 206
Table 12.3 Exploratory factor analysis—EMS 207

CHAPTER 1

Introduction

Susan L. Slocum

Craft Beverages and Tourism, Volume 1: The Rise of Breweries and Distilleries in the United States and *Craft Beverages and Tourism, Volume 2: Environmental, Societal, and Marketing Implications* are about the coming together of two significant industries: tourism and craft beverages. Kline & Bulla (2017) writes, "Craft beverage tourism is an exciting arena that intersects with many other current areas of growing scholarship, for example innovation and ingenuity, legislative oppression and globalization, and sense of place" (p. 2). While the geographic scope of these books is limited to the United States, their applicability is globally recognized as a means to better understand the implication of craft beverages in relation to destination development, experience development, entrepreneurship, marketing, social and environmental impacts, and consumer demand.

Volume 1 documented the significant rise in craft beverages as a means to differentiate corporate brands from the innovative craft brands that emerged within the market over the past decade. This phenomenon is attributed to changing regulations and the growth in celebrating a sense of place (Neister, 2008). Beer, cider, and distilled spirit production and

S.L. Slocum (✉)
Tourism and Event Management, George Mason University,
Manassas, VA, USA

consumption is on the rise internationally (Mathews & Patton, 2016) and is firmly rooted in expansions of the local foods movement, where artisan production is highly valued (Thurnell-Read, 2014). Volume 1 highlighted "the current practices and scope of research involving craft beer, cider and spirits as a substantial subset of craft and entrepreneurial small and medium enterprises that are emerging across the country" (Cavaliere, 2017, p. 173) and brought forth four basic themes: (a) the interdisciplinary of the craft beveragescape, (b) the evolution of the craft-turn in beverage production, (c) the role of co-creation for innovation, and (d) community redevelopment and sociocultural place making. The first volume highlights the potential of craft beverage production to be a leader in small business expansion which has the potential to contribute significantly to sustainable development in a variety of urban, suburban, and rural settings.

Volume 2 builds on the first volume by specifically tying craft beverages to the sustainability dialog and showcasing a variety of marketing implications within this growing niche industry. This volume employs a variety of approaches to situate sustainability, recognizing that sustainability is interpreted and implemented differently, not only within each industry but regionally as well (Jennings, 2009). Therefore, an appropriate definition of sustainability for this book would be "the balanced relationship of behavioral conditions that impact the environment, economics, and society in a way that still provides humans a viable present and future" (Hoalst-Pullen, Patterson, Mattord, & Vest, 2014, p. 111). Sustainability elements, such as resource reduction, social capital development, sense of place, and corporate social responsibility, are all examined in the following chapters. The focus on behaviors and practices in our sustainability discourse allows us to follow the emerging path that these craft entrepreneurs have traveled in their short history in the United States. We see that sustainability becomes a grassroots concept within a field that values "locality", "quality", and "craft" (Shortridge, 1996). The reliance on food tourism scholarly work (Murray & Kline, 2015) has neglected to highlight the subtle differences in how sustainability is expressed and showcased within the craft beverage industry. Therefore, this volume provides added insight into the specifics surrounding this emergent industry and research field.

However, the breadth and depth of sustainability practices in craft beverage tourism are variable. Cost saving is frequently a motivating factor in relations to energy, water, and landfill reductions. Resource conservation, such as "greening" the value chain, is readily apparent. Local sourcing and regional distribution patterns that reduce greenhouse gas emissions and support local economies are evidenced in these pages. Moreover, quality control is a primary motivation within local sourcing and sustainability practices. The "craft" aspect of craft beverages is the highest priority, where excellence and experimentation play the largest motivating factor in sustainability practices. There appears to be examples of fund raising for charities and support for other community endeavors. Celebrating local culture, and highlighting this in a variety of interpretation programs, is evident. Posters, signs, tours, and tasting all emphasize a sense of place tied to the crafters' locale. Yet, these practices are not consistent and vary, not only from region to region but also from business to business within the same geographical areas.

To that end, tourism studies have led us to believe that marketing craft beverages is also subtly different than general food. Within the sustainability dialog, educating the consumer is a primary objective (Bricker & Schultz, 2011), yet much of the sustainability accomplishments are surprisingly silent in these research chapters. Rather than focusing on the "green" initiatives commonly found in the craft beverage industry (reduced usage of water, electricity, and landfills), we find a deep sense of place that celebrates the "neolocal" of craft beverage production. Quality beer implies sustainable ingredients, and cost concerns push for "green" resource-reducing practices, yet the lack of self-celebrating messages that highlight these accomplishments is an important finding in these chapters. To that end, place-based principles, advertising, and interpretation are embraced by craft beverage producers over environmental conservation initiatives.

Jones begins the discussion on sustainability in Chapter 2 by evaluating corporate sustainability, where the reduction in resource use, commonly known as "greening", is more apparent in craft beverage circles than a more inclusive corporate social responsibility platform. He draws attention to what he references as the "opposite" of greenwashing by observing the lack of emphasis breweries place on showcasing their sustainability efforts. He claims five reasons for this: (a) obstacles to achieving third-party certifications, (b) disinterest from consumers and tourists, (c) the unusually cooperative culture of craft brewing, (d) combatting the elitist image of craft beer, and (e) avoiding sustainability comparisons with big beer. While acknowledging that social sustainability is much harder to quantify, therefore measure, Jones concludes that craft beverages have much more to offer within the sustainability dialog than is readily apparent.

In Chapter 3, Graefe, Mowen, and Graefe empirically examine craft beer consumers' travel and social world behaviors, their support of brewery localism, and the linkages between craft beer commitment, brewery neo-localism, and environmental attitudes and behaviors in Pennsylvania. In particular, their survey respondents did not always match the stereotype of the craft beer snob where beer is central to life or identity. They claim that customers prefer to visit breweries that use local products/ingredients and preclude that taste was the most important element in brewery visitation. However, the study does highlight a strong social world element among craft beer drinkers where consumers connect through craft beer blogs, social media, or brewery websites, and personal connections among craft beer enthusiasts are highly valued. Further, engagement with environmental activism is highly respected.

Lorr investigates water usage as a means to gauge sustainability efforts in three craft breweries in Michigan in Chapter 4. He finds that much of the problem relating to sustainable operations for craft beer tourism is the need to attract and retain a variety of customers, some of whom support, and others question, sustainability efforts. In particular, some local consumers are skeptical of rebates and tax incentives used by brewers to support water conservation and waste water treatment. He concludes that these breweries are advancing self-interest by using cost-saving mechanisms to reduce natural resource use as a means to better the environment and save money for their business. In turn, he finds that breweries are not moving forward in relation to social justice, community, or people, especially when attempting to remain apolitical.

Wright and Eaton's Chapter 5 returns to Michigan to explore the role that apple cider mills, structured as farm tourism destinations, play as mediators of rural cultural representation. They argue that the idyll rural representations have misled the visitor regarding rural realities, such as pervasive social, economic, and environmental challenges. These practices limit opportunities to engage urban populations in rural issues and propagate misunderstandings in relation to the modern realities of agricultural production. If agricultural tourism, in general, and cider mills, in particular, are encouraged as a means to economic prosperity in rural areas, there needs to be a more accurate representation that includes, rather than dissuades, conversations centered on rural truths. Only then can urban tourists become a better informed force for reshaping rural communities and the food system on which they are dependent.

Chapter 6 highlights the value of social capital, particularly bridging social capital in the sustainability dialog. Using data from both craft brewers and members of the tourism industry in Virginia, Slocum proposes that networks, feelings of trust and safety, reciprocity, participation, citizen power, common values and norms, and a sense of belonging necessitate social capital development. She discovers that small locally based tourism businesses, specifically bed and breakfast establishments, have the tightest social capital networks with local breweries. Large-scale tourism businesses, such as international hotel chains and tour companies, show the lowest social capital development with the craft beer industry. These findings are due to uneven power distribution, caused by differing access to high numbers of tourists. The involvement of the destination marketing organization shows promise in promoting craft beer tourism and breaking down mistrust in the industry.

Cavaliere and Albano discuss sense of place as it relates to craft spirit tourism marketing in Chapter 7. Using a modified version of the Multidisciplinary Framework for Place Conscious Education, they showcase craft spirit producers that are reinforcing and intentionally communicating sense of place in New Jersey. Through the use of farm names, a history of politics, sourcing of local ingredients, nostalgia, local geography, use of local folklore, and iconic infrastructure, stories of local heritage reverberate within the craft spirit industry. Unlike corporate conglomerates that push globalization, they emphasize how craft producers hold the power to reveal and recreate the "local" and argue that these entrepreneurs, particularly in New Jersey, are at the front line of capitalism. The locality provides a significant source of inspiration, which is further highlighted through the marketing they investigated, and demonstrates that local production can communicate the importance of place.

Understanding the demographic and physiographic makeup of craft beverage consumers is an important part of sustainability and holds many marketing implications. In Chapter 8, Curtis, Bosworth, and Slocum find that drink tourists to Utah tend to be middle-aged, highly educated, with few children at home, and highly involved in outdoor recreation and cultural activities. Drink tourism is a form of experiential consumption, and drink tourist motivations fall into four types of experiences: educational, esthetic, escapist, and entertainment. Additionally, lifestyle and value characteristics for drink tourists include an interest in local food experiences (shopping at farmers' markets and consumer-supported agriculture), preferences for organics, and participation in home-based activities such as cooking, gardening, and beer and wine making. These results indicate the importance of a healthy lifestyle, in terms of diet, and support for local farmers as a connection to drink production. It also highlights the association between the enjoyment of the environment and craft beverage consumption.

Cook compares three different Pennsylvania food festivals in Chapter 9. He finds four categories to express participants' feeling toward craft beverages: "only craft" that describes patrons who enjoy any type of macro and will only drink beer from craft breweries; "more craft than macro"; "more macro than craft"; and "craft only on special occasions", that is, patrons who rarely buy craft beer and may be first-time festival attendees. While most of the respondents in this study are men, the study finds that women enjoy beer, are experienced purchasers, and are likely to be a major part of the craft beer scene. Cook also concludes that beer festivals are not significant contributors to local economies, as visitors do not use hotels and restaurants and are generally local patrons (within 25 miles).

In addressing the urban craft beverage market in Chapter 10, Myles and Breen bring to light how collective action can facilitate cooperation even in the face of competition. Comparing three breweries, one in California, one in Kentucky, and one in the United Kingdom, they show how in cities, where beverage producers are located in close proximity to consumers, breweries are rooted in the "buy local" movement. These campaigns reinforce the message of freshness as a key component to quality beer, even if the ingredients within beer production are coming from different, sometimes distant, locations. Freshness of the product and local consumption are intertwined and the "localness" is crafted by the artisan. Myles and Breen reinforce the work of other authors in this text by recognizing that the production of craft drink can play an important role in place and identity formation, where artisan beverage producers, even in urban areas, can help to build local and regional identities.

In another urban area, Hayward and Battle's Chapter 11 helps to explain local industry identity dynamics through the interactions between the craft beverage industry, local residents, and tourists. The local industry articulates its image to residents and tourists in North Carolina and then adjusts its identity based on feedback from consumers. It is these interactions, they assert, that are critical to expressing local identity within the tourism and craft beverage industries. It is through intra-industry social networks, local trade institutions, tasting rooms, tours, and festivals that the stages for identity development evolve. In conclusion, Hayward and Battle encourage the use of coordinated messages that are balanced across industry stakeholders as a means to instill resiliency, which in turn brings a more sustainable form of beverage tourism that supports open conversations among the industry, residents, and tourists.

In Chapter 12, O'Neill and Sozen search for the psychological underpinnings of entrepreneurial motivation behind craft beer businesses. In a national study of craft beer owners, managers, and employees, they identify five components that motivate entrepreneurial activity in craft beer production: tax reduction and indirect benefits; need for approval; personal development; welfare and community considerations; and a need for independence. Since entrepreneurial engagement remains the backbone of economic development and a common policy objective of local governments, understanding small- and medium-sized business motivations can support the growth of the craft beverage sector. These results parallel prior general business research into the drivers of start-up companies; however, they find that craft beverage producers prioritize these motivation components differently. By determining the similarities and differences in craft beverage entrepreneurship, the authors hope to guide policy makers and investors interested in promoting food and beverage tourism development.

The editors bring these chapters to a conclusion in Chapter 13 by highlighting the common themes inherent throughout this book. These themes permeate throughout all the chapters and include quality sourcing and ingredients; resource reduction and efficiency; social sustainability; and the marketing of neolocal and sense of place. They also recognize that these two volumes only touch the surface of craft beverage tourism and encourage additional research. Utilizing an international lens to explore craft beverage tourism, the editors suggest cooperation versus competition, the political aspect of alcohol consumption and production, authenticity of craft production, bio-cultural conservation, and understanding the craft beverage tourist as areas of further inquiry.

The editors and contributors in this book series hope to provide a more theoretical foundation to support continued research in the area of craft beverage tourism. These volumes are just a start in the discourse on the many underpinnings and nuances of craft beverage production and consumption. While these volumes have focused primarily on the United States as a country still devolving from its temperance past and the cumbersome regulatory frameworks reminiscent of prohibition, we hope this text serves as a valuable resource of craft production in general, beverage production specifically, and for scholars worldwide. Furthermore, we hope that researchers continue to investigate niches within the food tourism field as a means to understand subtle differences in motivations to support the fruitful development of "foodie" destinations. Craft beverage is a growing contributor to positive place-based tourism development that holds the potential to offer sustainable options that celebrate the unique locality and heritage of rural, suburban, and urban communities.

REFERENCES

Bricker, K. S., & Schultz, J. (2011). Sustainable tourism in the USA: A comparative look at the global sustainable tourism criteria. *Recreation Research, 36*(3), 215–229.

Cavaliere, C. (2017). Conclusion. In C. Kline, S. L. Slocum, & C. T. Cavaliere (Eds.), *Craft beverages and tourism, volume 1: The rise of breweries and distilleries in the United States* (pp. 173–182). Switzerland: Palgrave Macmillian.

Hoalst-Pullen, N., Patterson, M. W., Mattord, R. A., & Vest, M. D. (2014). Sustainability trends in the regional craft beer industry. In M. Patterson & N. Hoalst-Pullen (Eds.), *Geography of beer* (pp. 109–118). New York, NY: Springer.

Jennings, G. (2009). Methodologies and methods. In T. Jamal & M. Robinson (Eds.), *The handbook of tourism studies* (pp. 672–692). Los Angeles, CA: Sage.

Kline, C., & Bulla, B. (2017). Introduction. In C. Kline, S. L. Slocum, & C. T. Cavaliere (Eds.), *Craft beverages and tourism, volume 1: The rise of breweries and distilleries in the United States* (pp. 1–10). Switzerland: Palgrave Palgrave Macmillian.

Mathews, A. J., & Patton, M. T. (2016). Exploring place marketing by American microbreweries: Neolocal expressions of ethnicity and race. *Journal of Cultural Geography*. doi:10.1080/08873631.2016.1145406.

Murray, A., & Kline, C. (2015). Rural tourism and the craft beer experience: Factors influencing brand loyalty in rural North Carolina, USA. *Journal of Sustainable Tourism, 23*(8–9), 1198–1216.

Neister, J. (2008). *Beer, tourism and regional identity: Relationships between beer and tourism in Yorkshire, England.* Unpublished master's thesis, The University of Waterloo, Ontario, Canada.

Shortridge, J. R. (1996). Keeping tabs on Kansas: Reflections on regionally based field study. *Journal of Cultural Geography, 16*(1), 5–16.

Thurnell-Read, T. (2014). Craft, tangibility and affect at work in the microbrewery. *Emotion, Space and Society, 13*, 46–54.

CHAPTER 2

Brewing Green: Sustainability in the Craft Beer Movement

Ellis Jones

INTRODUCTION

What drives this particular chapter is a burning need to answer the basic question, "who's the greenest of them all?" when it comes to craft beer. Like many craft beer enthusiasts, I have toured a good share of craft breweries and have enjoyed a wide selection of what the craft brewing world offers. However, unlike my experience with most other consumer product categories, I have never developed a sense of which beers were being produced in a more socially and environmentally responsible manner and which were not. While the research question has evolved into something slightly more sophisticated, along the lines of "what exactly does sustainability mean, in theory and practice, within the craft brewing industry?", the impetus behind the work remains largely the same. In essence, by the end of this chapter, the reader should be able to identify several of the greenest breweries in the United States and understand what exactly makes them "greener" than the rest.

E. Jones (✉)
Sociology and Anthropology, College of the Holy Cross,
Worcester, MA, USA

I begin this chapter by elaborating on the definitions of both corporate social responsibility (CSR) and one of its more recent offspring, corporate sustainability (CS), in order to provide a conceptual backdrop for the research project as a whole. I follow this with a description of how I combine social constructionism and grounded theory to create a research design that includes text analysis and interviews in order to explore how we make sense of what sustainability means in the world of craft brewing. From there, I discuss the inherent costs and benefits of the persistent environmental bias that has become ubiquitous to discussions of sustainability inside and outside of the business world. I then analyze the factors that contribute to the uniquely modest communication approach of craft breweries to touting their sustainability in a broader market landscape that is more often grappling with greenwashing. I end the chapter with a more theoretically oriented discussion of whether concepts like George Ritzer's (1983) *McDonaldization* and Jurgen Habermas' (1985) *colonization of the lifeworld* might offer us deeper insight as to how this shift from narratives of CSR to that of CS might indicate rough waters ahead for the sustainability efforts of craft breweries.

SUSTAINABILITY VERSUS SOCIAL RESPONSIBILITY: HOW COMPANIES APPROACH BUSINESS ETHICS

To better understand sustainability in the world of craft brewing, it is first necessary to comprehend the larger setting from which the term arose. CSR has been the standard term used to discuss the social and environmental practices of companies since the late 1960s (Dahlsrud, 2008). Other such terms include corporate accountability, responsible business, corporate citizenship, conscious capitalism, green business, and corporate social performance. While there is still no agreed-upon definition, CSR typically involves a triple bottom line analysis of company behavior that includes economic, environmental, and social impacts (Branco & Rodrigues, 2006; Kleine & Von Hauff, 2009; Taneja, Taneja, & Gupta, 2011) rather than just focusing on a purely economic bottom line (i.e. profitability). One of a handful of alternatives to CSR, CS, started gaining traction in the mid-1990s.

The origins of the term, CS, can be traced back to a broader conceptualization of sustainability coming out of the 1987 UN World Commission on Environmental Development's report entitled *Our Common Future* (aka *The Brundtland Report*), which defines the term simply as "meeting the needs of the present without compromising the ability of future generations to meet their own needs" (van Marrewijk, 2003, p. 101). This definition later becomes the basis of environmental impact assessment tools such as ecological footprints (Rees, 1992) and life-cycle analysis, which helps to shed light on why CS tends to maintain a largely environmental focus. While scholars have argued for decades that sustainability by definition includes social and economic (and perhaps even political and cultural) components of equal importance (Dyllick & Hockerts, 2002; Kleine & Von Hauff, 2009; van Marrewijk, 2003), the term stubbornly persists with its widely recognized, if troublingly narrow, environmental focus. A simple search on Google Scholar for the term "corporate sustainability" reveals that scholarly articles utilizing the concept of CS began to emerge in 1996 and have grown rapidly over the last two decades with 223 articles published in 2015 utilizing the term. As a point of comparison, utilizing the same procedure, we find that 2430 scholarly articles referring to CSR were published in 2015.

Seeing Beer Through Green Glasses: Researching Sustainable Breweries

In this study, I consider the concept of sustainability from a classic social constructionist perspective (Berger & Luckmann, 1967) in order to better grasp the dominant cultural narrative regarding the idea of sustainability in craft brewing. Because sustainability is still a fairly recent concept (and is particularly difficult for scholars to agree upon a common set of indicators to measure the sustainability of business), I contend that, at least initially, our most useful course of action is to understand how we (vis-à-vis news media) are socially constructing the concept of sustainable craft brewing. In order to do that, I start by investigating which breweries the news media are identifying as sustainable and then determine what seem to be the common set of narratives, practices, and indicators that ultimately lead news media to label these specific breweries as sustainable. To uncover these taken-for-granted notions of sustainability in the brewing world, I employ classical grounded theory (Glaser & Strauss, 1967) to orient my methodological approach. For this particular study, I utilize a two-step model combining (1) an initial text analysis of news media to identify sustainable breweries and (2) a series of personal interviews with sustainability coordinators at each brewery.

I began this work with a simple Google search of "sustainable breweries" and "sustainable beer" and collected news articles from the past five years that discuss the sustainability of multiple breweries nationwide (though I avoided self-published sustainability reports, brewery websites, and pieces focused on a single region or brewery). The resulting 26 articles included sources ranging from *The Huffington Post* and *Treehugger* to *Forbes* and *USA Today*.[1] Each article was coded for the number of times specific breweries were mentioned in order to assemble a rank list of breweries by number of articles that included their names when discussing the topic of sustainability. I also tracked the overall number of articles in which each brewery was mentioned with the aim of developing a better sense of how dominant certain breweries were in the current sustainable brewing narrative. I interviewed every brewery with more than five mentions in the article list,[2] including each of those listed in Table 2.1.

I interviewed the sustainability coordinators (or nearest equivalent) at the top eight breweries regarding how sustainability was conceptualized and practiced at each brewery. The 12 questions utilized in the interview schedule ranged from broad ("What does sustainability look like at your brewery?") to specific ("Do you work with any local farms or include any organic or fair trade ingredients?"), and each included one to three follow-up questions which were used when responses seemed to indicate richer data may be found through additional inquiries. Topics included social, environmental, and political impacts; what does and does not count as sustainability; communication efforts; unique examples of sustainability; characterizations of macrobrewers; greenwashing in the industry; other

Table 2.1 Rank list of sustainable breweries by mentions in online media

Rank	# of mentions (%)	Brewery	Location
1	23 (89)	New Belgium	Fort Collins, CO
2	18 (69)	Sierra Nevada	Chico, CA
3	15 (58)	Brooklyn	New York, NY
4	10 (39)	Great Lakes	Cleveland, OH
5	9 (35)	Brewery Vivant	Grand Rapids, MI
6	9 (35)	Odell	Fort Collins, CO
7	8 (31)	Full Sail	Hood River, OR
8	7 (27)	Alaskan	Juneau, AK
12	3 (12)	Allagash[a]	Portland, ME

[a]Allagash was chosen (after having been mentioned by name by some of the interviewees themselves) as a part of a methodological check to include breweries that may have otherwise been overlooked

brewers practicing sustainability; and the influence of customers and tourists. Each hour-long interview was conducted by phone and later transcribed and coded, organizing the data into the various themes of this chapter (Glaser & Strauss, 1967). The interviewees included five men and four women with titles ranging from the more official "sustainability specialist" and "people and planet coordinator" to simply the "brewmaster" and "maintenance engineer" identified as the point person to speak with about sustainability issues (see Table 2.2).

It should be noted at this point that I decided to reserve one interview slot for a brewery mentioned by the interviewees themselves as doing important work in the area of sustainability. In essence, this was a way of ensuring that I would not be missing any brewery that was particularly well-known within the craft brewery world for its record of sustainable behavior that was not being appropriately recognized by broader media. While several craft breweries were mentioned during the interviews themselves, the majority of them were covered on the original list of eight. Of the handful that were not on the original list, I chose to include Allagash Brewing after considering four factors: (1) it was one of the few breweries mentioned in more than one interview, (2) it had appeared three times in the online media search, (3) it added a measure of geographical diversity to the list, and (4) it was close enough to visit in person to gather on-site data.

One of the major weaknesses of this particular research methodology is its inability to capture and measure the sustainability work of smaller craft breweries. The list generated from major media coverage includes

Table 2.2 List of sustainable breweries ranked by size with position titles noted

Size rank	Brewery name	Barrels sold in 2014	Sustainability position title
3	Sierra Nevada	1,069,694	Sustainability specialist
4	New Belgium	945,367	Sustainability manager
11	Brooklyn	252,000	n/a
20	Alaskan	161,700	n/a
23	Great Lakes	149,948	Environmental programs coordinator
33	Full Sail	115,000	n/a
34	Odell	99,517	Sustainability coordinator
43	Allagash	70,406	n/a
a	Brewery Vivant	4780	People and planet person

[a]Not listed in the *Top 50 Breweries* list published annually by the *American Brewers Association*

only a single "small" craft brewer (Brewery Vivant) while the rest are large enough to be widely distributed and nationally recognizable by many, if not most, craft beer consumers and tourists. The top two breweries in the list (New Belgium and Sierra Nevada) are the third and fourth largest craft brewers in the United States (after Yuengling and Boston Beer), and the remainder of the breweries interviewed all fall within the 50 largest craft beer producers nationally (see Table 2.2). An interesting follow-up study might involve snowball sampling some of the smaller craft brewers (<10,000 barrels per year) that are recognized as sustainable by other brewers in their region of the country.

Brewing Sustainability: The Costs and Benefits of an Environmentally Focused Lens

Every brewery included in this study dedicates a significant portion of their website to discussing sustainable practices specifically, and two-thirds of the breweries produce annual sustainability reports. This focus on CS (and not CSR) is not insignificant in the case of craft brewing. As mentioned earlier, CS tends to focus primarily on environmental impacts while marginalizing or ignoring social impacts. However, what became apparent through the interviews is that the operationalization of sustainability has been narrowed even further to include mainly resource management concerns within the environmental arena: water, energy, and waste.

> So it's managing waste and managing energy use and managing water to make the product in the most efficient way and most responsibly as we can... They matter from not only an environmental standpoint but an efficiency standpoint and I think that, in the end, it's also good for business because you are just not wasting things.—Jamie, Full Sail

> We're looking at things like water use, energy use, transportation and fuel consumption, recycling and composting.—Cheri, Sierra Nevada

> We use a lot of resources from water to electricity to the ingredients that we put in our products to our packaging material. We want to try to minimize that impact, of course, for financial reasons and also because we want to have a good impact on our environment and our community. We're just trying to take on responsibly for what we do.—Corey, Odell

The Brewers Association publishes just three manuals on sustainability—one focused on water conservation, one focused on waste reduction, and one focused on saving energy (Brewers Association, 2014). An examination of the breweries' websites reinforces this more specific framing of sustainability. This narrow focus on resource management as the heart of sustainability in craft brewing comes with a particular set of benefits and costs.

The benefits of this approach are threefold. First, this type of resource management is easily measured (e.g. gallons of water reused, kilowatts of electricity saved, tons of spent grain repurposed) and thus improvements can be straightforwardly tracked from year to year. Second, these impacts can be communicated to employees, consumers, and tourists in a way that is easily comprehended and meaningful (e.g. 30% reduction in water consumption per bottle of beer produced over the past ten years). Third, in most cases, these kinds of efficiency improvements result in actual cost savings for the brewery over time; thus investments are recouped and can often lower the brewery's overall expenses in the short and long term.

However, from a sustainability perspective, this specific industry interpretation of the concept does not come without drawbacks. First, while easily quantifiable indicators of sustainability make for simpler tracking, improving, and communicating, an emphasis on such indicators typically means that those behaviors that are not as clearly calculable are given a lower priority, if not ignored altogether. Environmental impacts have a long history of having these general mathematical advantages over the less quantifiable impacts (human rights, humane animal treatment, discrimination, consumer empowerment, etc.). Second, environmental scholars often argue that efficiency upgrades, which can potentially lead to more profits in the medium to long term, may be perceived as "low hanging fruit" in the world of sustainable behaviors. Granted, the brewery upgrades are much more costly and achieve much higher efficiency gains; nevertheless, actions that ultimately pay for themselves (and lower expenses) may not be interpreted as evidence of deep environmental commitment as much as enlightened self-interest or even smart long-term investing. Finally, along these same lines, many brewers brought up the fact that the most efficient equipment often requires a larger scale operation that leaves many craft brewers without the upgrade option. Ironically, this means that the largest (mainstream) brewers are able to perform better on many, if not all, sustainability metrics than the craft brewers themselves (Schneider, 2014). This then allows dominant US macrobrewers to be "more sustainable" according to these numbers even if their motivations for doing so have little to do with environmental commitment but are instead largely or even exclusively due to the cost savings (i.e. profitability) involved. Some argue that the motivations do not matter if the results are the same, but others clearly see this as opening the door for mainstream brewers to greenwash their products by garnering credit for actions that they would take anyway due to their cost savings (Brennan & Binney, 2008; Panwar, Nybakk, Hansen, & Thompson, 2014).

Unpacking the Mystery of Quiet Sustainability: What Is the Opposite of Greenwashing?

Arguably, the most intriguing discovery to emerge from this research concerns the topic of *greenwashing*. Greenwashing is broadly defined as institutions making claims of environmental and social responsibility, generally for the purpose of marketing, that do not correspond with their actual practices on the ground (Lydenberg, 2002; Newson & Deegan, 2002; Owen & Swift, 2001). Greenwashing has become ubiquitous in most industries with estimates of this kind of ethically problematic form of advertising involving more than 90% of product and service categories (Derber, 2010). Despite asking directly about the problem of greenwashing in their industry, not a single interviewee could identify a case of greenwashing in the world of craft brewing. It became clear that there were no cases of greenwashing mentioned because, in large part, craft brewers are not actively marketing their sustainability efforts at all. It seems that craft brewers are avoiding the possibility of overselling their efforts in the area of sustainability by not making green claims to begin with.

> I don't think that [craft] brewers really do greenwash and I think the reason is that it's a bunch of privately owned companies who have ownership who are committed to sustainability and it's really a self-regulating environment.—Saul, Great Lakes

In addition, when I had an opportunity to visit one of the breweries (Allagash Brewing) as a tourist, I found no obvious signs of sustainability efforts on the 45-minute brewery tour (save for a few recycling bins being used by workers on the brew floor). While efficiency and resource conservation were briefly mentioned, they were a marginal component of the information provided, and the term "sustainability" was never uttered. In preparation for my visit, I investigated Allagash's website thoroughly and discovered that all of the items sold in their gift store are either "made in the US, organic, or made from recycled/reclaimed materials" (Allagash, 2016), but once in the store itself, there was not a single sign indicating this deliberately thoughtful focus on sustainability. I also found that the brewery was going so far as to have its branded merchandise manufactured by some of the most ethically reputable companies in their respective industries (often B Corp certified—a gold standard for CSR). Even the restroom area included waterless urinals and a water bottle refilling station, but there was no accompanying signage to explain their commitment to sustainability or the particular environmental impact these specialty appliances bring with them. Later, after methodically checking the labels on bottles and cans from each brewery (both in stores and via their online images), I found not a single mention of sustainability-related practices. In essence, outside of the website, there is little to no evidence to inform the typical customer or tourist that the brewery has any particular concern for environmental or social issues.

> We just are really striving to have a gentle approach with everything that we do and not hit people over the head with stuff.—Jeff, Allagash

> There are people that want to know [about sustainability] but we don't really want to hit people over the head with it.—Corey, Odell

> We've tried to be not too up in your face and present too much sustainability information.—Saul, Great Lakes

When I asked about this uncharacteristic approach in the interviews, what I heard repeatedly from each of the breweries was that they were pursuing sustainability because "it was the right thing to do" and that true responsibility was "doing the right thing when no one is looking". In other words, outside of a handful of pages and reports on the websites, there was almost no communication to indicate that these breweries were involved in some of the most highly regarded sustainability work in their industry. While some researchers have recently suggested that we label this form of undue modesty *brownwashing* (Kim & Lyon, 2014), I find the term troubling as it still suggests a certain amount of intentional "spin" when, in fact, what seems to be happening is simply that these brewers are not "receiving credit where credit is due". On the other hand, some may still argue that whether you are overselling or underselling your sustainability, both are problematic for the longer term goal of transparency in the industry that allows consumers to weigh all of the evidence for and against.

The question becomes, why would craft brewers be reluctant to communicate their sustainable practices to consumers and tourists who, all other things equal, are more likely to support more responsible companies? Not only is the market for ethical consumerism quite lucrative (Raynolds, Long, & Murray, 2014) but it would seem like an obvious way to establish a long-term relationship with consumers based on the integrity with which they address issues of sustainability. While the breweries themselves revealed no definitive answer to this question, I uncovered evidence of at least five potential factors that may each uniquely contribute to the explanation.

Third-Party Certifications The first factor involves additional hurdles for craft brewers to acquire the standard third-party certifications that have become recognizable markers of social and environmental responsibility in other industries, namely fair trade and organic certifications. Regarding fair trade, three of the four main ingredients for beer (grain, hops, and yeast) are typically sourced from North America or Europe, and the last, water, is obtained locally, thus eliminating most of the potential channels for fair trade products that are almost exclusively produced in the developing world (Stiglitz & Charlton, 2005). A number of brewers demonstrated interest in using organically certified ingredients but repeatedly mentioned three problems: a lack of adequate supply (Schneider, 2014), uneven ingredient quality, and significantly higher costs.

> We only do a little bit of organic here based on the cost difference and the quality of the ingredients and I think that is a big part of why you don't see organic used and marketed more in our industry ... the market demand isn't driving the cost down enough to make it feasible for everybody.—Kris, Brewery Vivant

Disinterest from Consumers and Tourists The second factor concerns a seeming disinterest from consumers and tourists in demanding sustainable options from brewers and rewarding those who provide those options at the checkout counter. In most consumer categories, demand for green or ethically sourced products is growing. However, my interviewees were not seeing the same demand in the world of craft beer.

> It's relatively hard to find organic ingredients; it's a chicken and egg situation, where the farmers aren't going to grow organic barley or organic hops without much demand ... the beer consumers haven't quite demanded that.—Matt, Brooklyn

> I think that the organic part of it is something that has interest to us, but we just haven't felt that there is a consumer demand for it for people who are interested in our product.—Andy, Alaskan

On the aforementioned tour of Allagash Brewing, not a single question was asked by tourists about sustainability. Most of the questions dealt with the mechanics of the process, the history, and the taste of their various beers. In the only example of its kind brought up by a brewery, New Belgium recounted how they had offered a certified organic beer for a number of years (*Mothership Wit*), but they ultimately discontinued it due to underperforming sales. Matt Gordon from Brooklyn Brewing speculated that craft beer checked so many other boxes for the average consumer (local, independent, artisanal, unique, etc.) that s/he does not need the additional reassurance of environmental or social responsibility.

> I think that when people view craft brewers, they get the small badge, local, independently-owned angle where the majority of them come from. For them, that sort of checks off a lot of boxes that are things that they would like to support. Organic would just be one more box on those other boxes to check.—Matt, Brooklyn

Even the two B-Certified companies in the group do not include the B Corp certification on their beer labels (B Corp certification is one of the most highly sought after, and difficult to achieve, markers of ethical behavior in any industry).

The Culture of Craft Brewing The third factor is related to the culture of craft brewing itself. Nearly half of the brewers interviewed at some point referred to the craft beer industry as cooperative rather than competitive. As I have learned, these kinds of comments are not just about a "rose colored glasses" approach to brewing but rather are a somewhat unique aspect of this business culture. "Collaboration beers" are an increasingly common sub-category of craft beers in which two or more breweries decide to work together on a particular specialty beer that they release for sale to the public. Smaller craft breweries are often built next door to or across the street from larger ones so that the latter can aid, assist, and even mentor the newcomers. Allagash, for example, is across the street from Foundation Brewing and Billings Bros. Brewery, and miles from almost anything else. In this microcultural climate, touting one's sustainability may be considered a shot across the environmental bow of other craft breweries, and thus they are content to improve their own practices behind the scenes, helping out fellow brewers whenever asked.

This ethic has also facilitated the creation and success of sustainability working groups at the national (Brewer's Association Sustainability Sub-Committee), state (Michigan Brewer's Sustainability Committee), and local (Ft. Collins Brew Water) levels.

Elitist Image Problem The fourth factor includes an elitist image problem that craft beer has been attempting to overcome since it entered the mainstream marketplace. Mainstream beer consumption in the United States has long been tied to a construction of masculinity that evokes hard work, toughness, sports, populism, and unfettered relaxation (Postman, Mystrom, Strate, & Weingartner, 1987; Stibbe, 2004). Craft beer, in contrast, is often tied to foodie culture, intellectualism, cosmopolitanism, and elitism. To a certain extent, craft beer has become the "wine" of beers. The most strident example of this cultural division can be seen in Budweiser's (2015) Superbowl television advertisement entitled "Brewed The Hard Way" in which "real American men" who drink Budweiser are contrasted with pretentious Euro-intellectual dilettantes drinking "effeminate" craft beer.

> There's only one Budweiser. It's brewed for drinking not dissecting. Let them sip their pumpkin peach ale. The people who drink our beer are people who like to drink beer brewed the hard way. (Budweiser, 2015)

Of course, as craft beer grows in popularity, the industry is attempting to shed this image (and seems to be largely succeeding). Sustainability labels may be seen as potentially thwarting this broader effort as social and environmental issues are still often perceived as something of an exclusively upper-middle-class or upper-class concern.

Big Beer The fifth and final factor concerns the role of Big Beer in sustainability. Craft breweries, for the most part, take a decidedly nonconfrontational approach to their larger competitors that at times even manifests as admiration (Jones & Harvey, 2017). Additionally, if sustainability is calculated strictly by the numbers (i.e. the most efficient use of resources), mainstream brewers' ready access to the latest technological innovations, via initial investment and scale, makes them look more sustainable on paper.

> A lot of the big brewers have been leaders in sustainable efforts. We have to say, for instance, on the CO_2 system that we were the first craft brewer to install a CO_2 recovery system in America, but that's because a lot of the big brewers had already done it. They were already recovering CO_2. They're

operating at such a massive level that they could see business efficiencies by recovering CO_2 and not paying an outside source for it.—Andy, Alaskan

Their electricity and water usage per barrel of beer produced is way lower than what most craft breweries can get to, in part because of how large they are. They can make big improvements with small changes.—Corey, Odell

From the limited part I know, I would say in some respect they are some of the most sustainable breweries because they are operating at such a large scale that a tiny bit more energy use or a tiny bit more water or a little bit more waste in their process or lower yields is huge dollars for them. In some respects, their financial goals will align with their sustainability goals.—Matt, Brooklyn

The combination of their amicable orientation to their larger competitors along with their inability to compete on purely quantitative benchmarks forms a significant disincentive for craft brewers to broadcast their sustainable behaviors. A sustainability marketing game which they are destined to lose when the numbers are scrutinized, albeit because of an artificially narrow focus on a handful of indicators, holds little appeal. It should be noted, though, that while no mentions were made of other craft brewers greenwashing their beer, there were a few mentions of potential greenwashing cases in the larger beer industry.

I recently saw an ad put together by Heineken that was basically promoting their green practices. Anything that pushes the bigger companies in that direction means that it holds us to higher standard, which is good, but I feel like we want to do it more for the right reasons or the sake of doing it whereas sometimes, from what I'm reading on a regular basis, the message does come off as 'We're doing this for marketing reasons.—Luke, Allagash

Furthermore, with their consistently strong emphasis on authenticity, craft brewers may not want to enter an arena (green marketing) inside of which their integrity may be scrutinized or even called into question in the first place—essentially a "better safe than sorry" approach to communicating sustainability practices.

Conclusion: The Big Picture of Sustainability in Craft Beer

Sustainability practices seem to be particularly well-rooted in the craft brewing world with two of the four largest craft beer producers modeling what best practices look like in this arena for the rest of the industry. Having said this, the version of sustainability being embraced leans strongly toward its most popular environmental interpretation and focuses most of its effort on the conservation of resources (water, waste, and energy). While social impacts (employees, community, philanthropy) and, to a lesser extent, political impacts (local legislation, national declarations) were mentioned in passing, environmental impacts are still the driving force behind the craft beer movement's sustainability achievements. This is not unexpected as it mirrors similar trends in other consumer industries (Dahlsrud, 2008).

The lack of communication around these sustainability practices, however, reveals craft brewers to be something of an anomaly. Rather than publicly emphasizing (or even overselling) their sustainable behaviors, craft beer producers seem quite satisfied engaging in these actions because it is "the right thing to do". At least five factors seem to be contributing to this norm in the industry: (1) obstacles to achieving third-party certifications, (2) disinterest from consumers and tourists, (3) the unusually cooperative culture of craft brewing, (4) combatting the elitist image of craft beer, and (5) avoiding sustainability comparisons with Big Beer.

Here we find, perhaps, the craft beer movement's most difficult conundrum: how to engage in their own sustainability efforts in the context of their relationship with the dominant US beer producers. In certain contexts, Big Beer is an ally on environmental issues (driving eco-technology innovations; participating in collaborative conservation efforts; and establishing benchmarks for water, waste, and energy efficiencies). However, due largely to advantages that can be linked directly to economies of scales, these dominant companies also achieve more measurable sustainability metrics than most, if not all, craft brewers can ever hope to attain. Additionally, these companies may be succeeding in implementing these efficiency gains purely for their financial benefits that can be realized more quickly and at more significant rates at the large-scale level. Craft brewers more laudable motives may be ultimately obscured when looking simply at the sustainability metrics in this case.

While the most intriguing empirical finding involves the tendency for craft brewers to practice sustainability in relative obscurity, the most interesting theoretical implication concerns the earlier mentioned, unanimous choice to pursue CS rather than CSR. Despite the advantages, it may be particularly important to consider that the growing popularity of CS, in the world of craft brewing and beyond, may be due to what Max Weber (1905), one of sociology's founding fathers, refers to as the process of *rationalization*, a progression within capitalism that strips much of society down to its most efficient aspects while discarding the rest. In some sense, CS can be understood as a minimalist version of CSR with most of the "irrational" components excised. Consider that under the CS model, sustainability is evaluated mainly by assessing resource conservation benchmarks (energy, waste, water) that are eminently quantifiable (number of gallons of water saved, tons of waste diverted, percentage of CO_2 emissions recaptured, and kilowatt-hours of electricity generated). At the same time, the much less quantifiable social components (community impact, human rights implications, social justice effects) are conveniently marginalized so that they are never seriously measured or evaluated, remaining merely responsible afterthoughts or "bonus points" if they are given consideration at all.

Perhaps an even more appropriate term can be found in George Ritzer's (1983) McDonaldization thesis (1983), derived from Weber's original conceptualizations. Ritzer points out that concerns with calculability, control, predictability, and efficiency drive social institutions to a kind of irrationality of rationality in which unintended consequences ultimately dwarf the intended ones. In this case, one could argue that CS is essentially the McDonaldization of CSR with aspects like social capital and human rights (inherently more challenging to measure and control) first marginalized, then ignored, and then forgotten in pursuit of the more easily quantified—if arguably less important—goals of waste reduction and resource management. If this is in fact what is occurring, then craft brewers should consider what Jurgen Habermas (1985) might call a re-embracing of the *lifeworld* to more effectively resist colonization by the *system*. Companies should consider how to integrate these messier, more qualitative, and more human-focused factors into their sustainability efforts, bringing them back into central focus along with existing environmental benchmarks so that they generate a more balanced approach to, and picture of, their achievements. This may also aid craft brewers, in particular, in their efforts to distinguish their own sustainability work from the practices of larger beer manufacturers, whose motivations and results may look significantly different under this new framework.

It seems that one of the major challenges craft beer producers face in the area of sustainability is how to distinguish their own laudable work in this area from what their larger competitors are offering. If one evaluates their sustainability strictly by the benchmarks and related metrics most commonly used in this area, Big Beer may be winning hands down. However, craft brewers have more to offer, benefits that are less easily quantified but arguably more important—things like commitment, creativity, integrity, diversity, and other similarly qualitative components. Additionally, we should not disregard the distinctive appeal that the notion of "local" holds for community members, locavores, cultural tourists, ethical consumers, and social activists. It may hold the key to translating sustainability efforts into a language that resonates with more of the public. In conclusion, while it may not be readily apparent on the bottle labels or even at the breweries themselves, a number of the most popular US craft brewers are engaging in innovative practices that are laying the foundation for the whole industry to evolve into a powerful force for environmental sustainability in the world of business.

NOTES

1. Sources include *Forbes, Intl Business Times, Treehugger, Triple Pundit, Huffington Post, Food & Wine, USA Today, Grist, Yes! Magazine, Green Business Bureau, Conservation Magazine, Inhabit, Opportunity Green, Paste Magazine, Chasing Green, Sustainable Planet, The Culturist, Lime Energy, MintPress News, Brew Bros, Where the Wild Grows, Web Ecoist, Eat Drink Better, Eco Sphere, Craft Brewing Business, Brewed for Thought,* and *Save On Energy.*
2. Ranked 9th is Eel River (five mentions), 10th is a tie between Bison and Stone (both at four mentions), 12th includes Allagash, Anderson Valley, Bell's, Boulevard, Deschutes, Fish, Lakefront, Red Hook, Steam Whistle, and Widmer (all at three mentions).

REFERENCES

Allagash. (2016). *Sustainability.* Retrieved January 2, 2016, from http://www.allagash.com/about/sustainability
Berger, P. L., & Luckmann, T. (1967). *The social construction of reality: A treatise in the sociology of knowledge.* New York: Anchor Books.
Branco, M. C., & Rodrigues, L. L. (2006). Corporate social responsibility and resource-based perspectives. *Journal of Business Ethics, 69*(2), 111–132.
Brennan, L., & Binney, W. (2008). *Is it green marketing, greenwash or hogwash? We need to know if we want to change things.* Partnerships, Proof and Practice – International Nonprofit and Social Marketing Conference 2008, University of Wollongong, July 2008.
Brewers Association. (2014). *Sustainability manuals.* Retrieved January 8, 2015, from https://www.brewersassociation.org/tag/sustainability-manuals
Budweiser. (2015, February 1). *Brewed the hard way: 2015 Superbowl commercial.* Television commercial. YouTube. Retrieved January 7, 2016, from https://www.youtube.com/watch?v=2uJKhkwTG64

Dahlsrud, A. (2008). How corporate social responsibility is defined: An analysis of 37 definitions. *Corporate Social Responsibility and Environmental Management, 15*(1), 1–13.

Derber, C. (2010). *Greed to green: Solving climate change and remaking the economy.* Boulder, CO: Paradigm.

Dyllick, T., & Hockerts, K. (2002). Beyond the business case for corporate sustainability. *Business Strategy and the Environment, 11,* 130–141.

Glaser, B. G., & Strauss, A. L. (1967). *The discovery of grounded theory: Strategies for qualitative research.* New York: Aldine De Gruyter.

Habermas, J. (1985). *The theory of communicative action: Lifeworld and system: A critique of functionalist reason* (Vol. 2). Boston: Beacon Press.

Jones, E., & Harvey, D. (2017). Ethical brews: New England, networked ecologies, and a new craft beer movement. In G. Chapman, J. Lellock, & C. D. Lippard (Eds.), *Untapped: Exploring the cultural dimensions of craft beer in the U.S.* Morgantown, WV: West Virginia University Press.

Kim, E., & Lyon, T. (2014). Greenwash vs. brownwash: Exaggeration and undue modesty in corporate sustainability disclosure. *Organization Science, 26*(3), 705–723.

Kleine, A., & Von Hauff, M. (2009). Sustainability-driven implementation of corporate social responsibility: Application of the integrative sustainability triangle. *Journal of Business Ethics, 85*(3), 517–533.

Lydenberg, S. D. (2002). Envisioning socially responsible investing. *Journal of Corporate Citizenship, 2002*(7), 57–77.

Newson, M., & Deegan, C. (2002). Global expectations and their association with corporate social disclosure practices in Australia, Singapore, and South Korea. *The International Journal of Accounting, 37*(2), 183–213.

Owen, D., & Swift, T. (2001). Introduction social accounting, reporting and auditing: Beyond the rhetoric? *Business Ethics: A European Review, 10*(1), 4–8.

Panwar, R., Nybakk, P. K., Hansen, E., & Thompson, D. (2014). The legitimacy of CSR actions of publicly traded companies versus family-owned companies. *Journal of Business Ethics, 125*(3), 481–496.

Postman, N., Mystrom, C., Strate, R., & Weingartner, C. (1987). *Myths, men & beer: An analysis of beer commercials on broadcast television.* Falls Church, VA: AAA Foundation for Traffic Safety.

Raynolds, L. T., Long, M. A., & Murray, D. L. (2014). Regulating corporate responsibility in the American market: A comparative analysis of voluntary certifications. *Competition & Change, 18*(2), 91–110.

Rees, W. E. (1992). Ecological footprints and appropriated carrying capacity: What urban economics leaves out. *Environment and Urbanization, 4*(2), 121–130.

Ritzer, G. (1983). The "McDonaldization" of society. *Journal of American Culture, 6*(1), 100–107.

Schneider, F. (2014). *Identity creator, political food and gourmet drink: An approach to craft beer culture in Bristol.* Unpublished master's thesis, University of Gastronomic Sciences IT.

Stibbe, A. (2004). Health and the social construction of masculinity in Men's Health magazine. *Men and Masculinities, 7*(1), 31–51.

Stiglitz, J. E., & Charlton, A. (2005). *Fair trade for all: How trade can promote development.* New York: Oxford University Press.

Taneja, S. S., Taneja, P. K., & Gupta, R. K. (2011). Researches in corporate social responsibility: A review of shifting focus, paradigms, and methodologies. *Journal of Business Ethics, 101*(3), 343–364.

van Marrewijk, M. (2003). Concepts and definitions of CSR and corporate sustainability: Between agency and communion. *Journal of Business Ethics, 44*(2–3), 95–105.

Weber, M. (1905). *The Protestant ethic and the spirit of capitalism.* New York: Scribner's.

CHAPTER 3

Craft Beer Enthusiasts' Support for Neolocalism and Environmental Causes

David Graefe, Andrew Mowen, and Alan Graefe

INTRODUCTION

Consumption of and societal interest in craft beer in the USA have grown dramatically over the last two decades. While varying opinions abound concerning the perfect definition of craft beer and craft brewers, the Brewers Association (2016a) suggests that three criteria should be considered: craft brewers are small (annual production of six million barrels or less), independent (less than 25% of the craft brewery is owned by a member of the beverage alcohol industry that is not itself a craft brewer),

---

D. Graefe (✉)
Department of Natural Resources & the Environment, Marshall University, Huntington, WV, USA

A. Mowen
Recreation, Parks and Tourism Management, The Pennsylvania State University, University Park, PA, USA

A. Graefe
The Pennsylvania State University, University Park, PA, USA

and, perhaps most importantly, traditional (a majority of beers produced derive their flavors from traditional or innovative brewing ingredients and their fermentation). The number of microbreweries (i.e., companies that produce craft beer) had grown to 2768 by 2013 (Brewers Association, 2016b), and with an annual growth rate of 10%, craft beer sales are projected to represent nearly 15% of the total beer industry by 2020 (Demeter Group, 2013). Craft beer is now firmly established as a vibrant segment of the US beverage industry and is a frequent topic of conversation among business leaders, community development organizations, academicians, and even politicians.

Given that microbreweries are steeped in local community/culture, some have attributed their growth and popularity to be reflective of a larger societal shift toward localism (e.g., support for local goods/services, local branding) and environmental sustainability (Flack, 1997; Schnell & Reese, 2003). Increasingly, it seems that craft breweries are not only aligning/branding their products with local names, images, and history, but are also active partners in local and environmental causes and organizations. Among many similar examples, Yellowstone Valley Brewing of Billings, Montana partnered with the Yellowstone River Conservation District to produce a craft beer (692 No Dam Brew) to celebrate Yellowstone River as the longest free-flowing river in the lower 48 states. Sales proceeds from this craft beer went directly toward conservation projects on the river.

Despite increased attention devoted to the craft beer industry, few studies have assessed the characteristics, attitudes, and behaviors of craft beer consumers. Watson (2013) explained that around the turn of the century, craft beer consumers were characterized as being middle-aged, predominantly white, male, and having relatively high incomes. However, he suggested that craft beer consumers have become more diverse in recent years, with a growth in consumption among young women, Hispanics, and those with lower incomes. Murray and O'Neill (2012) found that home-brewers could be characterized as being predominantly male, middle-aged, and having high incomes and education levels.

In addition to these and other demographic profiles (e.g., Mintel Business Market Research Report, 2012), craft beer consumers have been further examined as the focus of several outdoor recreation and tourism marketing campaigns (Howlett, 2013; VisitBend.com, 2016). Indeed, tourism promoters (e.g., convention and visitor bureaus) and environmental non-profit park organizations (e.g., park conservancies) are making subtle and not-so-subtle associations between their region/organization and craft beer in order to stimulate tourism and promote environmental engagement. For example, Visit Corvallis promotional spots encourage visitors to bike their trail systems and stop off for a craft beer after the ride. VisitBend's promotional spots target craft beer enthusiasts more explicitly by comparing their zeal for outdoor adventure and taste in craft beer against the relaxed lifestyles of those who want lime in their beer. Not all craft beer media portrayals are favorable, however, and craft beer enthusiasts have recently been mocked or parodied in popular culture (Goldfarb, 2013; Tuttle, 2016). Despite these developments, we know very little about consumers' craft beer-related behaviors, particularly those linked to craft beer travel and social worlds (e.g., brewery trips, pub memberships, social media activity). Current discussions suggest craft beer consumers are strongly supportive of brewery localism (Mathews & Patton, 2016; Murray & Kline, 2015) and environmental causes. The prevalence of these attitudes and behaviors among craft beer consumers, however, has received less scientific inquiry.

Psychological involvement with a product, brand, or leisure activity can contribute to attitudes and behaviors consistent with images and norms associated with that subject (Kyle, Absher, Norman, Hammitt, & Jodice, 2007). Previous examinations of craft beer market trends, brand loyalty, and motivations for craft beer affiliation have provided valuable insights regarding peoples' tendency to purchase products that endorse unique local communities and environments. However, they provide little evidence of precise social-psychological pathways or connections between consumers' psychological involvement with craft beer and subsequent attitudes and behaviors supportive of brewery localism and environmental stewardship. Are consumers who are more psychologically involved with craft beer more likely to express favorable attitudes toward brewery neolocalism and are they more likely to express pro-environmental values/behaviors than their less involved counterparts? If they are, such enthusiasts could represent a promising constituency for agencies and non-profits to promote local tourism, community causes, events, and environmental activism through craft beer. We contend that, because of craft breweries' historic reliance on localism and their alignment with environmental causes as part of their business model, committed consumers who report craft beer as an important part of their identities and social worlds will be more likely to favor brewery neolocalism and will themselves report more pro-environmental values and behaviors than individuals with a more casual (or less committed) relationship with craft beer. The precise nature and pathways of these specific relationships have not been empirically examined, until now. This study assesses craft beer consumers' travel and social world behaviors, their support of brewery localism, and the linkages between craft beer commitment, brewery neolocalism, and environmental attitudes and behaviors. A brief discussion of the craft beer industry's ties to neolocalism and sustainability follows.

The Craft Beer Industry, Localism, and Sustainability

The US craft beer industry has experienced remarkable growth over the last few decades, and this growth is expected to continue in the years to come. However, this growth follows a significant decrease during the preceding decades. Flack (1997) explained that the years following the Second World War saw a substantial decrease in the overall number of American breweries. This decrease can be attributed to consolidation, centralization, and fierce competitive tactics by the largest American breweries, which drove out hundreds of smaller breweries. According to the Brewers Association (2016b), the total number of breweries in the USA decreased from 476 in 1945 to only 92 in 1980. However, the years following this period saw substantial growth, and in 2014, there were a total of 3464 US breweries, which represents a nearly 4000% total growth rate between 1980 and 2014. While the overall beer industry has experienced a slight decline in recent years (partially due to an upsurge in wine and spirit sales), the American craft beer segment continues to experience considerable growth (Brewers Association, 2016c; Demeter Group, 2013). Many scholars have suggested that this growth is due not only to changing tastes but also to changing values among US citizens.

The relationship between craft breweries and local cultures has been well documented in the research literature and can be seen in countless anecdotal examples (we encourage you to take a trip to your local microbrewery and see for yourself). Flack (1997) argued that although many Americans have become more sophisticated and passionate about the taste of their beverages (and perhaps dissatisfied with the lack of flavor and variety of mass-produced beer), their demands do not require microbreweries, as a wide variety of imports are readily available. Why then has the US craft beer industry grown so dramatically? Flack argued that while imported beers may satisfy increasingly diverse and sensitive palates, they do not satisfy the neolocal craving (i.e., the feeling of belongingness to a unique local community, along with the rejection of global, national, or even regional popular culture and modernization). Domestic craft breweries, on the other hand, may satisfy this craving for affiliation with unique local identities and cultures. The recent boom in the US microbrewery industry is at least partially due to citizens' desires to connect with their local identities and cultures, and to develop a local sense of place. Flack (1997) wrote, "Unlike most pop culture phenomena, microbreweries engender a strong, self-conscious attachment to their localities" (p. 49). Further, the neolocalism movement represents a conscious attempt to break free from the "smothering homogeneity of popular, national culture" (Schnell & Reese, 2003, p. 46). This connection can be seen in countless examples where craft brewery brands, varieties, and activities are chosen to highlight and popularize the idiosyncratic characteristics of local communities. For example, Otto's Pub and Brewery in Pennsylvania has chosen names for its beer varieties that reflect specific local natural resources and recreation areas (e.g., Spring Creek Lager, Slab Cabin IPA). Further, many microbreweries make a conscious effort to contribute to local economies and health by committing to the use of local products and organic ingredients. Some microbreweries have even begun to form partnerships with local organizations to promote community health and sustainability. For example, Fathead Brewery conducts several fundraising events to contribute to the Cleveland Metroparks Trails Fund (Cleveland Area Mountain Bike Association, 2013).

While the neolocalism movement may be partially responsible for craft breweries' engagement with local communities, another contributing factor is likely related to the public's awareness of and engagement with environmental conservation and sustainability initiatives. Prior research has shown that consumers' relationships with foods and beverages can shape or be shaped by their personal environmental worldviews and environmental behaviors. Ethical consumerism has been discussed since at least the early 1970s and has become a central component of business planning and marketing. Traditionally, markets were most often segmented on the basis of demographic and behavioral variables. However, in the 1970s, researchers began to recognize the existence and importance of the "socially conscious consumer" and began to realize the potential of market segmentation based on personality and socio-psychological attributes (Anderson & Cunningham, 1972; Webster, 1975). The proportion of the population that fits this description appears to have grown dramatically, and this market segment is now most commonly referred to as Lifestyles of Health and Sustainability, or LOHAS.

LOHAS has become a prominent market segment that is relevant to a wide range of industries and activities. French and Rogers (2010) provided a nice summary for readers who are unfamiliar with this terminology:

> It (LOHAS) refers to a wide range of industries, corporate activities and products/services that are designed to be environmentally conscious, sustainable, socially responsible, and/or healthier—both for people and the planet. The LOHAS consumer, in particular, is the leading-edge portion of the population that is attracted by their belief systems and values and who make their purchase decisions with these criteria in mind. LOHAS consumers are also used as predictors of upcoming trends, as they are early adopters of many attitudinal and behavioral dynamics. (p. 1)

Although it is difficult to determine the exact theoretical underpinnings of the LOHAS market segment, some suggest it stems from Ray and Anderson's (2000) and Aburdene's (2007) writings concerning the cultural creative class and conscious capitalism. LOHAS and other related market segments are characterized as a hybrid lifestyle—a form of "as well as" postmodern ethics. This lifestyle is characterized by seven attributes: realistic and spiritual, affinity toward technology and nature, health and pleasure, individual but not exclusive, discerning by no status luxury, modern and value-driven, and self-centered but interested in community welfare (Kreeb et al., 2009).

LOHAS consumers and NATURALITES (another large market segment that is concerned with health and sustainability but more concerned with personal health than planetary health) may be particularly attracted to establishments and events that are focused on craft beer and brewing and especially those that are committed to sustainable practices, health, and locally grown products. Research conducted by the Natural Marketing Institute (2008) provides support for this proposition. In 2007, the LOHAS segment represented 19% of the US adult population and the NATURALITE segment represented another 19%. Use of both organic and natural foods increased significantly between 2003 and 2007 in the general population. LOHAS consumers are by far the heaviest users of organic and natural foods, with 60% of households among this segment reporting natural food/beverage purchases in 2007 (National Marketing Institute, 2008). Though healthy and sustainable food and beverage choices are becoming more mainstream, the LOHAS market segment remains a frontrunner in these purchasing behaviors. Further, general population consumers rated artisanal foods and beverages as most important among a list of purchasing criteria including fair trade, organic, and local. This suggests a nationwide trend away from generic, mass-produced products toward foods and beverages that are unique, special, different, created using traditional methods and provide experiential satisfaction.

Though this is a nationwide trend, the LOHAS segment rated the importance of artisanal foods higher than any other market segment. Interestingly, locally sourced foods and beverages were ranked fourth among the aforementioned purchasing criteria (i.e., after artisanal, fair trade, and organic). Moreover, National Marketing Institute's researchers expect this issue to become increasingly important as consumers continue to become more aware of the benefits of freshness and the impact of food miles (i.e., the distance that food/beverages travel from their origins to their destinations) on the environment. Most of the research documenting this trend has focused on food systems and providers (e.g., farmers' markets, community-supported agriculture networks, community gardens, etc.), yet the recent emergence and success of the craft beverage industry suggests it may also play a significant role in promoting neolocalism and environmental engagement among its patrons.

In a study of Pennsylvania craft beer patrons, we test this idea by examining craft beer consumers' attitudes toward brewery neolocalism (e.g., brewery local sourcing and cause activities) as well as their environmental attitudes and behaviors. We also assess whether levels of craft beer involvement or commitment contribute to support for brewery localism and environmental attitudes/behaviors. Finally, to provide important market information on the craft beer consumer (as opposed to the industry), we assess a number of craft beer travel and social world behaviors.

Study Methods

Sampling and Data Collection

This study employed a survey research design to gather information from visitors to Pennsylvania microbreweries, craft brew pubs (i.e., bars that primarily serve craft beer), and craft beer festivals/events. Patrons to such establishments and events were approached using a roving intercept technique and were asked to complete an anonymous, self-administered questionnaire designed to ascertain their attitudes and behaviors concerning craft beers/breweries, their perceptions toward brewery localism, and their general environmental beliefs and behaviors. A systematic sampling technique was employed, in which every "nth" patron (based on the number of possible subjects) was approached. Those who agreed to participate in the survey were compensated with a small bottle-opener key chain. Patrons who appeared to be intoxicated were not asked to complete the questionnaire.

Researchers visited 11 microbreweries, brew pubs, or craft beer festivals to obtain data for this study. Although the data collection procedures resulted in a convenience sample, an effort was made to collect data from a variety of regions within Pennsylvania. This was deemed the most appropriate and feasible sampling technique, as the purpose of this study was to explore relationships between individuals' commitment to craft beer and their attitudes and behaviors relating to localism and the environment, rather than providing a representative sample of all Pennsylvania craft beer drinkers.

Variables and Analyses

Several variables were included on the questionnaire to gather information about patrons' behavioral and psychological attributes relating to craft beer, localism, and the environment. Multiple items were included to measure behavioral commitment to craft beer (e.g., craft beer-related visitation, tourism behaviors, expenditures, and engagement in social worlds). Psychological commitment to craft beer was measured using a four-dimensional enduring leisure involvement scale, which has been conceptualized in terms of "personal relevance", and reflects the degree to which people devote themselves to an activity or product (Kyle et al., 2007). Environmental attitudes were measured using the New Ecological Paradigm (NEP) scale, which has become the most widely used measure of environmental concern worldwide (Dunlap, 2008). The NEP was developed, at least in part, as a means to measure and document the evolution of environmental worldviews among humans and has been used to characterize individuals' attitudes as reflecting either a traditional, utilitarian philosophy (i.e., the dominant worldview) or a more ecological philosophy. Personal engagement in pro-environmental behaviors (i.e., behaviors that support environmental protection, stewardship, or sustainability) was measured using a seven-item index adapted from Theodori and Luloff (2002). To develop our measure of support for sustainable brewing practices and localism, we referenced magazine articles and blogs and interviewed brewery operators to develop a range of brewery localism items. These items involved local sourcing of food/ingredients, brewery involvement in cause activities, sustainability practices, and opposition to taste as the sole criterion for beer preference.

Factor analysis of the sustainability and neolocalism item pool suggested three factors with good reliability (local sourcing, brewery cause activities, and taste only). These three factors were used to represent support for craft brewery neolocalism (see Table 3.4). In addition to computing descriptive statistics, a path analysis was conducted to examine the influence of psychological commitment to craft beer on (1) support for brewery localism, (2) environmental attitudes, and ultimately (3) pro-environmental behaviors.

Results

A total of 306 questionnaires were collected, with a response rate of 85%. A majority of study participants were male (71%), with an average age of 40 years. Thirty-three percent reported household incomes less than $50,000 per year, while 27% reported household incomes of $110,000 or more per year.

Participants made an average of about 30 trips (or visits) to craft breweries/brewpubs over the past 12 months (Table 3.1). Over half (59%) reported they visited at least once per month and 16% visited an average of once per week. On average, participants visited five different craft

Table 3.1 Craft beer travel and visitation patterns

Travel behavior	Mean	Median	SD
Number of total trips to craft breweries or brewpubs (that brew their own beer) within the last 12 months	29.6	10	53.9
Number of different craft breweries or brewpubs visited in the last 12 months	5.1	4	5.6
Number of beer festivals or "meet the brewer" events attended in the last 12 months	2.2	1	4.3
Do you seek out and visit local microbrew pubs during your work or vacation travels? (percent yes)	79.0		

breweries over the last 12 months. Fifteen percent visited ten or more different breweries over that time period. Participants attended an average of two beer festivals or meet-the-brewer events over the last 12 months. Thirty-six percent did not attend any, while 23% attended three or more festivals/events. Within the past 12 months, participants made an average of two trips of 50 miles or more specifically to visit a brewpub or to visit a brewpub as part of another trip. Fifty-six percent said that they took at least one such trip while 15% reported five or more trips. More generally, 79% of study participants stated that they seek out and visit local microbrew pubs during their work or vacation travels.

In terms of beer "media-related" activities, 28% subscribed to or read a craft beer magazine, 39% "liked" a brewery or brewpub on Facebook, 28% "followed" a brewery or brewpub on Twitter or through an e-mail list, and 69% had visited a craft brew blog, website, or brewery home page on the web over the last 12 months (Table 3.2). Twenty-seven percent were members of a pub/mug club or similar membership club at a microbrewery.

Measures of craft beer commitment/involvement were divided into four domains and computed into indices following previous research (Kyle et al., 2007). The item composition of the commitment domains and summary results are shown in Table 3.3. Participants reported the strongest agreement (mean = 3.69) with the first commitment domain, attraction, showing a relatively high importance of craft beer to them personally. The second strongest element of commitment was the social bonding domain (mean = 3.20), containing items related to social interactions involving

Table 3.2 Craft beer "social worlds" behaviors

Social worlds behavior	% yes
Over the past 12 months, have you visited a craft brew blog, website, or brewery home page on the internet?	68.8
Within the past 30 days, have you struck up a conversation with a stranger about beer at a microbrew pub?	65.4
Over the past 12 months, have you "liked" a brewery or brewpub on Facebook?	38.9
Over the past 12 months, have you "followed" a brewery or brewpub on Twitter or through an email list?	27.9
Over the past 12 months, have you subscribed to or read a craft beer magazine?	27.8
Are you a member of a pub club, mug club, or have similar membership at a microbrewery?	27.1

Table 3.3 Craft beer commitment/involvement

Commitment item/domain/index[a]	Mean	SD	% agree	CITC	α if deleted
Drinking craft beer is one of the most enjoyable things I do	3.65	1.02	61.7	0.791	0.794
I have little or no interest in drinking craft beer (reversed)	4.42	0.83	89.7	0.554	0.885
Drinking craft beer is one of the most satisfying things I do	3.28	1.01	40.7	0.780	0.799
Drinking craft beer is important to me	3.43	0.99	53.8	0.741	0.815
Domain: Attraction	**3.69**	**0.81**	–	α = 0.865	
A lot of my life is organized around drinking craft beer	2.49	1.07	16.1	0.802	0.888
Drinking craft beer occupies a central role in my life	2.31	1.09	14.0	0.858	0.868
I invest most of my energy and resources in drinking craft beer	2.02	1.09	6.3	0.821	0.883
I try to structure my weekly routine around drinking craft beer	2.18	1.05	11.0	0.737	0.910
Domain: Centrality	**2.24**	**0.93**	–	α = 0.913	
I enjoy discussing craft beer drinking with my friends	3.48	1.14	62.6	0.680	0.861
Most of my friends are in some way connected with drinking craft beer	3.05	1.18	39.7	0.688	0.860
Drinking craft beer provides me with an opportunity to be with friends	3.57	1.02	63.5	0.699	0.859
Special people in my life are associated with craft beer drinking	3.22	1.15	45.7	0.690	0.859
I prefer to be around others who share my interest in craft beer drinking	3.08	1.07	35.4	0.731	0.853
I identify with the images associated with craft beer drinking	2.75	1.06	23.6	0.649	0.866
Domain: Social bonding	**3.20**	**0.88**	–	α = 0.880	
When I drink craft beer, I can really be myself	2.78	1.08	22.9	0.681	0.917
When I'm drinking craft beer, I don't have to be concerned with the way I look and behave	2.54	1.12	18.9	0.658	0.920
My true self emerges when I drink craft beer	2.39	1.08	14.1	0.847	0.900
You can tell a lot about a person by seeing them drink craft beer	2.54	1.11	20.8	0.700	0.915
To a large extent, drinking craft beer provides one of the few outlets where I can be myself	2.26	1.04	11.4	0.807	0.905
Drinking craft beer says a lot about who I am	2.39	1.07	15.0	0.768	0.908
Drinking craft beer allows me to express myself	2.46	1.10	18.7	0.838	0.901
Domain: Self-expression	**2.47**	**0.89**	–	α = 0.922	
Overall commitment index	**2.87**	**0.75**	–	α = 0.950	

Corrected item total correlation (CITC)
[a]Measured on a five-point scale (from 1 = strongly disagree to 5 = strongly agree)

craft beer. Somewhat less important were items within the self-expression domain (mean = 2.47). The respondents were less likely to feel that drinking craft beer "allows me to express myself" or "says a lot about who I am". Finally, items within the centrality domain received the lowest scores (mean = 2.24), suggesting that most craft beer drinkers do not organize their lives around drinking craft beer or invest a lot of energy and resources to it.

As noted earlier, the measures of support for brewery neolocalism and sustainability practices were original to this study and fell into three domains (Table 3.4). Respondents showed strong support for both the local sourcing (mean = 3.74) and brewery cause activities (mean = 3.84) domains. The majority of craft beer drinkers were more likely to visit

Table 3.4 Support for brewery neolocalism

Neolocalism item/domain[a]	Mean	SD	% agree	CITC	α if deleted
I am more likely to visit restaurants/bars that use locally grown meat or produce in their menu	3.82	1.10	70.2	0.765	0.785
I am more likely to visit microbrew pubs that use local ingredients in their beer	3.66	1.04	62.1	0.720	0.827
I am more likely to select menu items that use locally grown products	3.74	1.02	63.9	0.738	0.811
Neolocalism domain: Local sourcing	3.74	0.93	–	α = 0.863	
I like to support brewpubs that are actively involved in local environmental causes	3.79	1.02	64.3	0.643	0.835
I like to support brewpubs that recycle their brewing materials	3.83	0.99	64.1	0.729	0.798
I like to support brewpubs that sponsor or support outdoor recreation clubs or groups	3.69	1.01	60.2	0.643	0.835
Craft breweries should do all that they can to operate in a sustainable manner	4.05	0.89	76.0	0.593	0.852
Neolocalism domain: Brewery cause activities	3.84	0.82	–	α = 0.853	
I don't care about how "green" a microbrewery/pub is as long as their beer is tasty	2.75	1.23	31.0	–	–
I don't care what products are used in making the beer, as long as it tastes good	3.00	1.30	39.6	–	–
Neolocalism domain: Taste only	2.87	1.09	–	α = 0.663	

[a]Measured on a five-point scale (from 1 = strongly disagree to 5 = strongly agree)

establishments that use locally grown products in their beer and menu items. Likewise, most of the participants were more supportive of brewpubs that act sustainably (e.g., recycle their brewing materials) and support environmental and outdoor recreation-related causes. Respondents were more likely to disagree with the statement, "I don't care about how 'green' a microbrewery/pub is as long as their beer is tasty".

The environmental attitude and behavior variables were measured with instruments reported in previous studies. Environmental attitudes were measured by the 15-item NEP scale (see Dunlap, 2008). The composite index score represents people's general views on the environment or level of environmental concern. Personal engagement in pro-environmental behaviors was assessed with a seven-item index adapted from Theodori and Luloff (2002). These items included significant actions or efforts by individuals to support the environment or environmental causes (Table 3.5). Of these seven behavioral items, craft beer drinkers were most likely to "stop buying a product if it causes environmental problems" (mean = 2.39)

Table 3.5 Pro-environmental behaviors

Behavioral item/index[a]	Mean	SD	% frequently or always	CITC	α if deleted
I contribute money or time to an environmental or wildlife conservation group	1.98	0.80	21.0	0.637	0.825
I will stop buying a product if it causes environmental problems	2.39	0.82	37.3	0.530	0.841
I attend public hearings or meetings about the environment	1.37	0.65	7.4	0.631	0.828
I contact a government agency to get information or complain about an environmental problem	1.37	0.65	7.1	0.602	0.832
I read conservation or environmental magazines, blogs, or newsletters	1.94	0.90	21.7	0.712	0.812
I watch television specials on the environment	2.33	0.85	38.1	0.563	0.836
I vote for or against a political candidate because of his/her position on the environment	2.29	0.95	37.0	0.636	0.826
Pro-environmental behavior index	**1.93**	**0.57**	–	\multicolumn{2}{c}{α = 0.850}

[a]Measured on a five-point scale (from 1 = never to 5 = always)

or "watch television specials on the environment" (mean = 2.33). They were less likely to report more highly engaged behaviors such as "attending public hearings or meetings about the environment" (mean = 1.37) or "contacting a government agency to get information or complain about an environmental problem" (mean = 1.37).

The variables described above were analyzed in a path analysis using multiple regression with environmental activism (measured by pro-environmental behaviors) as the dependent variable. Craft beer commitment was the independent variable, and support for brewery neolocalism/sustainability and environmental attitudes (measured by the NEP scale) were tested as mediating variables for the relationship between craft beer commitment and pro-environmental behaviors (Fig. 3.1). Results showed that craft beer commitment was significantly related to all three domains of support for brewery localism. Commitment most strongly affected support for brewery cause activities (R^2 = 0.13) followed by support for local sourcing (R^2 = 0.09) and the taste-only domain (R^2 = 0.03). Participants with greater craft beer commitment showed more support for brewery cause activities and local sourcing and were less likely to not care about these issues as long as their beer tastes good. While these relationships were statistically significant, craft beer commitment was not a powerful predictor of neolocalism domains, as shown by the relatively low R^2 values. The three neolocalism domains in turn influenced environmental attitudes, with support for brewery cause activities and local sourcing again showing a positive influence and the taste-only measure showing an inverse relationship. Finally, both the environmental attitudes and support for brewery neolocalism predicted pro-environmental behaviors, with the environmental attitudes measure (NEP) showing the greatest influence (β = 0.375). The environmental attitudes and support for neolocalism together accounted for 40% of the variance in pro-environmental behaviors among craft beer drinkers.

The mediation tests showed that the effects of craft beer commitment on environmental attitudes and behavior were fully mediated by the three domains of support for brewery neolocalism, as the effect of commitment was no longer significant when the neolocalism domains were included in the analysis. Environmental attitudes partially mediated the effects of support for brewery neolocalism, as the effects of these domains remained significant, though were smaller, when the NEP scale was included in the analysis.

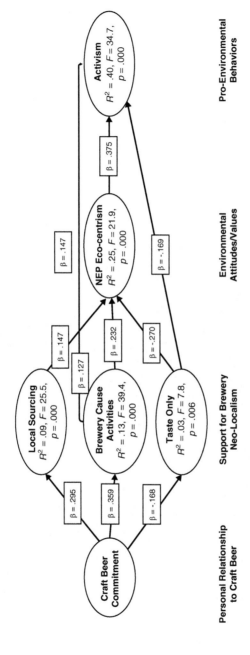

Fig. 3.1 Craft beer commitment, neolocalism support, and environmentalism

Discussion and Conclusion

The rapid and sustained growth of the craft beer industry over the past decade has been remarkable. While much attention has been devoted to documenting this growth, its roots, and in debating the future of craft beer, less attention has been paid to documenting the attitudes and behaviors of craft beer consumers themselves, their commitment to craft beer and neolocalism, and their attitudes and behaviors with regard to environmentalism. The growing attention being paid to humans' relationship with foods/beverages and how this relationship affects our environmental worldviews and activities warrants more research on craft beer consumers and their environmental views. Is it true that craft beer consumers are more likely to support breweries that embrace local sourcing and support environmental causes/collaborations? Emerging craft beer-centric tourism promotion campaigns (e.g., VisitBend) as well as brewery sponsorship of (and support for) environmental organizations suggest that it is. Will such perceptions and relationships be evident among a sample of craft beer enthusiasts? We surveyed a sample of craft beer drinkers at various venues across Pennsylvania to find out.

Results from our survey were supportive of much of the current rhetoric tied to craft beer neolocalism. However, respondent behaviors and attitudes did not always match the stereotype of the ultra-committed craft beer snob (e.g., Goldfarb, 2013; Tuttle, 2016). For example, respondents reported visiting, on average, five different craft breweries, two beer festivals, and making 30 total trips to craft breweries over a 12-month period. These numbers are hardly the picture of a hyper-committed consumer, suggesting that craft beer is a part of, but not central to, consumers' daily lives. Despite these moderate behaviors, a majority of our respondents (79%) indicated that they sought out craft beer venues during their work or vacation travels. These results are supportive of convention and visitor bureau (CVB) efforts to include craft breweries in their tourism marketing and branding (as discussed by Howlett, 2013). While trendsetter CVBs such as Visit Corvallis and VisitBend have been linking their promotions and events to craft beer enthusiasts for years, other CVBs could consider including craft breweries as part of their local food/sourcing promotions, particularly those businesses with a track record of environmental engagement and partnerships.

Beyond consumption and travel behaviors, our results indicated a strong social worlds element among craft beer drinkers. A majority of our respondents reported connecting with craft beer blogs, websites, or brewery websites. Their use of social media and membership in craft beer clubs was less frequent (under 30%). However, as our study was conducted in 2012, we only assessed a limited range of social media (e.g., Facebook and Twitter) and this could have influenced these data.

The personal connections among craft beer enthusiasts—both in terms of close friendships and in terms of casual exchanges with strangers—were very evident in the data. For example, over 60% reported conversing with a stranger at a brewery about craft beer. Moreover, of all the psychological commitment domains, social bonding (among friends) was among the strongest. In this sense, craft beer seems to provide a source of affiliation and socialization through mutual respect and love for craft beer. This finding builds upon the results of Murray and Kline (2015), who found that a connection with the community was the most important of several factors influencing brand loyalty to rural craft breweries. Again, CVBs, festivals, and local organizations who seek to attract craft beer enthusiasts should consider ways to provide programs/venues for both "thick" and "thin" social interactions among close friends as well as strangers who share the common bond of craft beer. For example, ice-breaking craft beer games and activities might be offered to promote social interaction at events, while promoting a kinship among attendees who are relative strangers bound by a common interest. Bonding capital could be enhanced by providing group tours and reunion events around craft beer.

Beyond social bonding, other psychological domains of craft beer commitment (e.g., centrality to life, self-expression) were not as strong. Our respondents generally rejected the notion that their life centered around craft beer or that it was a means to reflect their personal identities. These particular findings may be subject to social-desirability bias, as admitting that "one structures their life around craft beer" could be seen as a sign of a problem with alcohol.

A central argument posed in this chapter (and by other craft beer scholars) is that craft beer is tightly bound to neolocalism and environmental values and behaviors (Flack, 1997; Mathews & Patton, 2016; Schnell & Reese, 2003). Our empirical results generally support these propositions. We found craft beer drinkers liked to visit breweries that used local products/ingredients in their business and rejected the notion that taste was the most important element that dictates their support for a particular brewery. Not only did local sourcing matter but so did craft breweries' engagement with environmental activism. A majority of respondents indicated that they liked to support craft brewpubs that are actively engaged with or sponsor local outdoor and environmental organizations and causes (as also discussed by Howlett, 2013). Our findings suggest that new craft breweries wishing to establish consumer loyalty consider neolocalism, both philosophically and in their activities. The Highland Brewing Company in Asheville, North Carolina, is an exemplar of these connections. Proceeds from release of their seasonal Clawhammer Oktoberfest beer support the conservation efforts of the Southern Appalachian Highlands Conservancy and the US Fish and Wildlife Service. Many other craft breweries engage in regular sponsorship or donations to local environmental causes and these connections should continue to be made by local CVBs and nature/environmental conservation organizations.

The environmental values and behaviors of craft beer drinkers were also assessed in this study. Do consumers with higher levels of psychological commitment to craft beer and higher levels of support for brewery neolocalism also report more pro-environmental attitudes and behaviors? In our study, we analyzed a series of path models and found direct and indirect connections between these variables. However, the notion that the personal relationship or commitment to craft beer is directly linked to more eco-centric values and more frequent environmental behaviors was not evident in our study. Rather, commitment was positively related to consumer support for brewery neolocalism, which in turn was directly and indirectly related to environmental attitudes and behaviors. While commitment to craft beer is an indicator of multiple neolocalism domains, it does not directly manifest environmentalism. What does this mean on a practical level? Perhaps one argument to increase environmentalism among craft beer drinkers is to elevate the importance and support of brewery localism—another would be to make more direct connections between craft beer itself and environmental causes. Regardless, our data presents a strong case that localism is an important element of brewery support among brewpub patrons and is directly tied to environmental causes. Natural resource agencies, tourism marketers, conservancies, and other related organizations should consider both craft beer enthusiasts and craft breweries as important constituents and partners in meeting their environmental missions.

In this study, 40% of the variation in pro-environmental behaviors was explained by environmental attitudes and neolocalism domains, which is quite respectable given the results of previous research. Thapa (1999) explained that researchers have consistently found statistically significant but relatively weak relationships between environmental attitudes and behaviors, and suggested that while people like to consider themselves as environmentalists, such perceptions likely do not have a strong influence on personal behaviors that support the environment. Our results provided slightly stronger support for the influence of environmental attitudes on pro-environmental behaviors. However, there are still opportunities to improve the power of statistical models attempting to explain environmental behavior. Future researchers should consider additional variables that have been shown to influence human behaviors. For example, the Theory of Planned Behavior posits that behaviors are influenced by a variety of factors, including attitudes, subjective norms, perceived behavioral control, and behavioral intentions (Ajzen, 1991). The incorporation of such variables would likely increase the predictive power of models attempting to explain pro-environmental behaviors.

REFERENCES

Aburdene, P. (2007). *Megatrends 2010: The rise of conscious capitalism.* Charlottesville, VA: Hampton Roads Publishing Company Inc.

Ajzen, I. (1991). The theory of planned behavior. *Organizational Behavior and Human Decision Processes, 50,* 179–211.

Anderson Jr., W. T., & Cunningham, W. H. (1972). The socially conscious consumer. *Journal of Marketing, 36,* 23–31.

Brewers Association. (2016a). *Craft brewer defined.* Retrieved from https://www.brewersassociation.org/statistics/craft-brewer-defined/

Brewers Association. (2016b). *Number of breweries.* Retrieved from https://www.brewersassociation.org/statistics/number-of-breweries/

Brewers Association. (2016c). *National beer sales and production data.* Retrieved from http://www.brewersassociation.org/press-releases/brewers-association-announces-2013-craft-brewer-growth/

Cleveland Area Mountain Bike Association. (2013). *Brew party at Fathead's tasting room.* Retrieved from http://camba.us/wp/archives/4841

Demeter Group Investment Bank. (2013). *State of the craft beer industry: 2013.* Retrieved from http://demetergroup.net/sites/default/files/news/attachment/State-of-the-Craft-Beer-Industry-2013.pdf

Dunlap, R. E. (2008). The new environmental paradigm scale: From marginality to worldwide use. *Journal of Environmental Education, 40*(1), 3–18.

Flack, W. (1997). American microbreweries and neolocalism: "Ale-ing" for a sense of place. *Journal of Cultural Geography, 16*(2), 37–53.

French, S., & Rogers, G. (2010). *Understanding the LOHAS consumer: The rise of ethical consumerism.* LOHAS Online. Retrieved from http://www.lohas.com/Lohas-Consumer

Goldfarb, A. (2013, October 24). *The most mockable things beer snobs do*. Retrieved from http://www.esquire.com/food-drink/drinks/a25340/beer-snob-mockery-1013/

Howlett, S. (2013). *Bureaus and beer: Promoting brewery tourism in Colorado*. Retrieved from http://citeseerx.ist.psu.edu/viewdoc/download?doi=10.1.1.6 83.4739&rep=rep1&type=pdf

Kreeb, M., Motzer, M., & Schulz, W. F. (2009). LOHAS als Trendsetter für das nachhaltigkeitsmarketing. In C. Schwender et al. (Eds.), *Medialisierung der nachhaltigkeit. das forschungsprojekt balance(f): Emotionen und ecotainment in den massenmedien* (pp. 303–314). Marburg: Metropolis Verlag für Ökonomie, Gesellschaft und Politik.

Kyle, G., Absher, J., Norman, W., Hammitt, W., & Jodice, L. (2007). A modified involvement scale. *Leisure Studies, 26*(4), 399–427.

Mathews, A. J., & Patton, M. T. (2016). Exploring place marketing by American microbreweries: Neolocal expressions of ethnicity and race. *Journal of Cultural Geography*, 1–35.

Mintel Group Ltd. (2012). *Market research report: Craft beer – U.S.* Retrieved from http://store.mintel.com/craft-beer-us-november-2012

Murray, A., & Kline, C. (2015). Rural tourism and the craft beer experience: Factors influencing brand loyalty in rural North Carolina, USA. *Journal of Sustainable Tourism, 23*(8–9), 1198–1216.

Murray, D. W., & O'Neill, M. A. (2012). Craft beer: Penetrating a niche market. *British Food Journal, 114*(7), 899–909.

Natural Marketing Institute (NMI). (2008, March). *Understanding the LOHAS market report: Consumer insights into the role of sustainability, health, the environment and social responsibility*. Retrieved from http://andeeknutson.com/studies/LOHAS/General%20Health%20and%20Wellness/11_LOHAS_Whole_Foods_Version.pdf

Ray, P., & Anderson, R. (2000). *The cultural creatives: How 50 million people are changing the world*. New York: Three Rivers Press.

Schnell, S. M., & Reese, J. F. (2003). Microbreweries as tools of local identity. *Journal of Cultural Geography, 21*(1), 45–69.

Thapa, B. (1999). Environmentalism: The relation of environmental attitudes and environmentally responsible behaviors among undergraduate students. *Bulletin of Science, Technology, and Society, 19*(5), 432–444.

Theodori, G. L., & Luloff, A. E. (2002). Position on environmental issues and engagement in proenvironmental behaviors. *Society and Natural Resources, 15*, 471–482.

Tuttle, B. (2016, February 6). *Budweiser doubles down by mocking craft beer again in super bowl ad*. Retrieved from http://time.com/money/4210344/budweiser-super-bowl-50-ad-mock-craft-beer/

VisitBend. (2016). *The Bend Ale Trail™ – Your beer adventure begins in Bend, Oregon*. Retrieved from http://www.visitbend.com/Bend_Oregon_Activities_Recreation/Bend-Ale-Trail/

Watson, B. (2013). The demographics of craft beer lovers. Brewers Association.

Webster Jr., F. E. (1975). Determining the characteristics of the socially conscious consumer. *The Journal of Consumer Research, 2*(3), 188–119.

CHAPTER 4

Pure Michigan Beer? Tourism, Craft Breweries, and Sustainability

Michael J. Lorr

Introduction

Water is a natural resource beer drinkers take for granted. Some craft beer fans have rudimentary knowledge of how to brew beer and its intense water use. But it is unlikely they know how their favorite brewery attempts to address sustainability, especially as they pertain to water. Water is the most important ingredient in beer. The quality of the water greatly affects the quality of the beer. Water, like many natural resources, is taken for granted by the American public until something is drastically wrong. Recently, the Flint Michigan water crisis, and a more broadly aging water delivery system, put many cities at risk of lead exposure. The aging Enbridge oil pipelines under the Mackinac Bridge in the Great Lakes are overdue for a catastrophe that would affect the water supply of all human settlements in the area, affect tourism, and the ability of area breweries to access suitable water to make beer. Some brewers have started a "No Fracking Way" clean water craft beer initiative to highlight environmental problems. Brewing near one of the more abundant supplies of freshwater on the planet, Lake Michigan, some craft breweries address the water issue

M.J. Lorr (✉)
Aquinas College, Grand Rapids, MI, USA

because they understand how business, tourism, and environmental sustainability are interrelated.

This chapter describes and explores the relationships between the popularity of craft beer, impacts on water resources, and broader sustainability goals. Limited research exists regarding how craft brewing impacts the environment or how craft breweries address sustainability and water concerns. Accordingly, this chapter attempts to fill that gap by using the West Michigan region as a case study focused on the sustainability efforts of three West Michigan breweries: Founders, Short's, and Brewery Vivant.

Craft Beer in the USA, West Michigan, and Grand Rapids

Writing on beer abounds, much of this is popular travel writing about frequenting breweries (Canham-Nelson, 2012) or retracing the steps of making popular brews, like India Pale Ale (Brown, 2009). These writers also dabble in popular histories of beer and breweries (Brown, 2006; Neu, 2011), beer's social significance (Brown, 2003; Rosenzweig, 1991), or beer encyclopedias (Alworth, 2015; Oliver, 2012). There are also publications about styles and tastes of beers (Jackson & Lucas, 2000). Many beer consumers want to know the style of beer they consume—from its particular ingredients down to the type and origin of yeast used in the fermentation (Oliver, 2012). The core of this literature is that people like craft beer. In 2014, beer consumption per capita in the USA was 75.8 liters or 20.0 gallons ("List", n.d.). In 2014, the overall beer market was valued at $101.5 billion, and craft beer makes up $19.6 billion or 19% of that market (Nodrum, 2015).

Grand Rapids, Michigan, using the public relations and tourism moniker, "Beer City, USA", is a small city of about 200,000 in the Midwest that is capitalizing on this growing beer bubble (Pure Michigan, 2013). The larger West Michigan region is home to more than 40 breweries (Ruschmann & Nasiatka, 2006). One recent Grand Rapids tourism promotion is a stampable brewery passport and a free T-shirt for those who visit 8 of the 23 participating breweries (Pure Michigan, 2015).

Water Pollution and Brewing

Beer making is a water intensive process. MillerCoors is attempting to reduce their water use to 3.4 barrels of water for 1 barrel of beer, whereas many home brewers use 5 gallons of water to make 1 gallon of home brew (Alworth, 2015; Oliver, 2012). At the average commercial brewery, it takes 224 gallons of water to produce 32 gallons of beer, or 7:1, and many in the beer industry are attempting to reduce that to 3:1 (Alworth, 2015; Oliver, 2012). Many large breweries come close to this ratio by doing maintenance and upkeep and adding a sustainable process or piece of equipment. In the brewing process, water is used for sanitation, so the beers ferment properly, and for consumer-safety reasons. Breweries in West Michigan are situated in a water oasis near the Great Lakes. Yet, Michigan breweries must act as stewards for the water they are dependent on. It seems that a lot of the time, it is a combination of pressure from Clean Water Laws, Municipal Water, and a desire to be environmentally aware as a business or as a group of individual employees working within a business. Kouwenhoven's *The Beer Can by the Highway* (1961) highlights the tensions between abundance and waste as a manifestation of what has shaped the USA, the world, and the future. The jarring image of people wantonly throwing beer bottles and cans out of their car windows underscores the reality and tragedy of the Anthropocene—the impact humans have on the planet, its species, and its resources (Kolbert, 2014).

Sustainability and Craft Beer Tourism

The twenty-first century is cited as a time when humans are pursuing sustainable development and creating livable cities while at the same time presiding over massive extinction events (Agyeman, Bullard, & Evans, 2003; Kolbert, 2014; Whitehead, 2007). Sustainable societies and the businesses in them strive to respect the "triple bottom line": bettering the environment, improving the economy, and alleviating social injustice. Chouinard and Staley (2012) and others who write about the potentials of corporate social responsibility argue that businesses can be responsible in a genuine way by addressing people, planet, profit, and ultimately steering society toward better environmental, economic, and equitable or socially just conditions. Averill (2016) reported that brewers are society's most creative conservationists—from designing mass marketable rubber bricks for reducing water use in toilet tanks to recycling gray water (wastewater) into beer by using NASA-derived technology. These innovations are not widely known by the public or publicized by the industry.

A primary reality of the craft brewing industry is competition. In order to compete, these breweries need to have a compelling product to attract and retain new customers and are only really able to do this when they grow large enough to become a destination brewery, such as Founders for example. As environmental policy and planning researchers, Agyeman et al. (2003, p. 2) succinctly summarize, "[sustainability requires] a better quality of life for all, now and into the future, in a just and equitable manner, whilst living within the means of supporting ecosystems". Until definitions of sustainability and business expand to directly emphasize social and environmental justice, sustainability remains a rhetoric that provides environmental amenities to wealthier green-oriented consumers. Acknowledging this, how substantive are breweries' attempts at sustainability and how capable are breweries at implementing policies to deliver on broad green promises and specific water pollution and conservation solutions?

A small portion of the Pure Michigan tourism advertising campaign is aimed at getting people to visit the state's craft breweries. Mowforth and Munt (2016), and other scholars, question whether tourism and sustainability substantively address environmental, economic, and equity issues or the extent to which sustainability is used as a marketing strategy to encourage mass tourism. Examples like the Enbridge pipeline and Flint water issues call to question the contradictions of the Pure Michigan tourism promotion; still, this tagline encourages people to alter their everyday life—to potentially "beer the change" as Brewery Vivant claims below. Living greener lifestyles becomes an ongoing tactic, strategy, and process of people changing everyday life, in the hopes of pushing the limitations of contemporary material and policy infrastructures, but also being bound by those limits (Lorr, 2012a, 2012b). Brewers wanting a sustainable industry in Michigan attempt to make sustainable changes to their business and to educate their consumers, and some consumers attempt to buy local sustainable beer. There is little evidence suggesting that these green innovations lead toward a more sustainable craft beer industry and more sustainable-minded consumers. To the extent that there is a business and lifestyle change, what are the implications of these heavily market-oriented approaches?

METHODS

In order to address breweries' attempts at water sustainability and the ability of brewers to deliver on their green promises, this research offers interview responses from three West Michigan breweries' sustainability officers and employees. Those involved were tasked with implementing sustainability policies in these workplaces. This chapter aims to provide a descriptive and exploratory study of these three West Michigan breweries and how they address water use and broader environmental impacts of brewing.

Research Sites: The Breweries

Grand Rapids, Michigan, is home to Founders Brewery, a popular craft brewery started in 1997. Recently, Founders entered into a deal with a brewery from Spain, illustrating their trajectory toward Big Brewer, away from craft. Short's Brewing in Bellaire is a mid-sized brewer in Northwest Michigan. Short's original plan was to make and distribute beer only within Michigan. After January 2016, they changed that plan so they could distribute out of state and continue to grow. Grand Rapids, Michigan, is also home to Brewery Vivant. Vivant is a much smaller brewery specializing in Belgian-style beers and their business model is as a Certified B corporation brewery restaurant. In the USA, these "benefit" corporations are a type of for-profit entity certified by the non-profit B lab to meet standards of social and environmental performance, accountability, and transparency. This also includes positive impacts on society, workers, the community, and the environment in addition to profit as legally defined goals. Benefit corporations differ from traditional C corporations in purpose, accountability, and transparency, but not in taxation. This means that every aspect of Brewery Vivant's brewing, restaurant, and business overall is dedicated to sustainability goals and metrics.

These three breweries depend on the abundant supply of water in Michigan and have had to address their water use and the amount of "less than pure water" that results as a byproduct from their brewing. Preliminary research on these sites revealed that all of these breweries make at least some public relations commitment to sustainability. Perhaps Brewery Vivant, because it is a Certified B corporation, has made a more substantive commitment to institutionalizing all aspects of sustainability, but Short's has its own water treatment facility that rivals the municipal public works in Bellaire, and Founders has worked well with the local city government in Grand Rapids to pay appropriate fees for the water it uses and puts back into the system as waste. These breweries are also members of the Michigan Brewers Guild and its subcommittee on sustainability.

Interview Method and Limitations of This Study

After hundreds of informal interviews with brewers, pub staff, craft beer enthusiasts, and others related to the industry, I formally interviewed at least one sustainability officer/employee from each brewery. These interviews took place between March and June in 2016 and were intended to highlight sustainability initiatives at each brewery. In order to do this, I emailed each brewery my interview schedule, which asked broadly about how sustainability, craft brewing, and tourism interrelate. I requested both written responses and the ability to call, interview, and ask follow-up questions, which resulted in three original respondents, who are the official sustainability officers at the three breweries and via snowball sampling included seven more respondents who were subordinates or sustainability interns at the breweries— ten total respondents. This study is limited by the geographic location of the breweries in the West Michigan region, the preliminary and exploratory nature of the study, and the time constraints for this study. Per sociological research ethics, Institutional Review Board approval, and ethnographic research conventions, individuals had knowledge of possible publication of their responses, granted consent, and were given the opportunity to read the first full draft of this chapter to offer suggestions and responses. In an effort for transparency, replicability, and accountability, these are real businesses and the first names of real people. This preliminary research provides the basis for a larger research project on brewing, sustainability, and tourism.

FINDINGS

This section reports on what brewery sustainability employees said about how their breweries "do" sustainability, starting with Founders, Short's, and ending with Brewery Vivant. I ordered the findings in this way because of brewery size and also because of my impressions from the interviews of the spectrum of rationales for "doing" sustainability at each brewery. While all of these breweries are "doing" sustainability in some way, it is clear that Brewery Vivant has sustainability at the core of its business strategy because it is a Certified B corporation. By comparison, Founders "does" sustainability as an ancillary component of their business model. Founders' main purpose is to provide high-quality beer, their green marketing and public relations remarks are on message, they meet the minimums required by law (as in the case of water treatment), and they work to stay on the right side of the progressive city, its local customers, and the plentiful beer tourists. These breweries stated that making quality beer was their number one job and made the argument that to make high-quality beer, they need access to high-quality ingredients. In order for that access to continue, these breweries acknowledged that they need to play a role in addressing many aspects of the current environmental crisis, still there was a spectrum of concern on which Founders was on one end and Brewery Vivant was at the other end.

Founder's Interview Data

Founders does not define sustainability on its website yet, but it does have a sustainability archive. The most recent post is from 2014. Founders' Strikeforce Green, their internal committee that is working on implementing a sustainability vision and mission for the brewery, created a hard-to-find poster outlining Founders' commitment:

> Vision: to act as a driving force and resource to develop and implement practices and procedures that are cost effective, feasible, and impactful. Mission: to promote a philosophy of environmental stewardship tantamount with world-class beer production through engaging and empowering the Founders family and community to reduce waste and increase efficiency with focus on water, energy, and solid waste.

Brett, Founders Cellar Manager, and Liz, their Sustainability Intern, said Strikeforce Green, Founders' "green think tank", came up with the above mission, vision, and goals of sustainability. They reiterated that making quality beer first makes good business sense and is the primary concern. Founders tracks detailed Brewers Association determined sustainability metrics in the above poster. Brett said:

> A lot of breweries are around the 10 barrels of water to 1 barrel of beer ratio—the best are closer to around 4:1. Our goal at Founders is 3.6:1 but it's hard because ever since we've opened we've been in a state of perpetual expansion, it's always a learning curve and it's steep.

Brett and Liz asserted that sustainability is always questionable and problematic because of its upfront cost. The example they gave was the 6–7 figure cost of solar panels and installation. Because of this upfront cost,

Founders decided to focus on education first to get buy-in from their staff. This "low hanging fruit" approach teaches those at the brewery how to recycle and what to recycle. Brett said that most sustainability work has to be incentivized:

> There needs to be incentives for this type of work and there aren't all that many. We really exist to meet customer demand and we are distanced from the more political aspects of a lot of the environmental and sustainability movements. Our approach is to get buy-in and be accountable to our patrons so that we are responsible and make sure people have access to our product.

Founders may be able to do more if there were more incentives for sustainability. Until those incentives materialize, they partner with local environmental non-profits like West Michigan Environmental Action Council (WMEAC) and they have an annual sustainability report. But they continue to come back to trying to make the highest quality beer with the highest quality ingredients. Impressionistically, this seems like a way to say, while sustainability and being green is popular, they will at least join the club. When questioned further, they say sustainability is a part of their "better ingredients" argument, but it is not a primary interest for marketing or otherwise, even though what they are doing is innovative and important. Brett said:

> We work with the city to save on effluent surcharges and utilize sidestreaming as much as we possibly can. It's a learning experience with the city and both sides are constantly in negotiation with each other. We've become more efficient and have less of a drain on the local water systems than we used to. We can't quite be like Bell's Brewery in southwest Michigan and their water treatment facilities because we have limited space in this urban location. Anaerobic and aerobic water treatment would face resistance for smell, sight, and location. An efficient brewery makes for a better community. Liz as an intern is the first step to solidifying sustainability and to institutionalize it further over the next two to three years by creating a sustainability department like in the brewers guild and cultivating more community connections like the kind we have with the city, WMEAC, and GR whitewater to make the Grand River a rapids again.

Founders is taking incentivized steps toward sustainability. The sustainability intern position was a volunteer job when their StrikeForce Green committee started. When I first interviewed the Founders respondents in June of 2016, the sustainability intern position was a paid part-time add-on role to serving. In their response to the first draft of this study, Founders recently made the sustainability intern position a full-time position—"sustainability coordinator". Founders is heading toward more robust sustainability efforts as a part of their current expansion. If they had more incentives coupled with regulatory targets, sustainable brewing may become the norm.

Short's Interview Data

Short's brewery and its culture concerning sustainability come from its location in the outdoorsy up North. Short's defines sustainability on their website in this way:

> The environmental definition of sustainability is "the quality of not being harmful to the environment or depleting natural resources and thereby supporting long-term ecological balance," which sounds like something we should all be doing and there are a few breweries right here in Michigan doing their part. Short's Brewing Company is one of those breweries that has taken massive action in the last couple years focusing on sustainable growth, and they are not alone.

This definition is similar to other breweries and Short's has similar robust energy, recycling, and water programs. The interview I had with Tyler, Short's Director of Quality and unofficial Director of Sustainability, focused on water use and sustainability efforts. For comparison, Brewery Vivant does 3000 or 4000 barrels of beer a year, Short's does 38,000, and Founders does around 160,000 barrels. Short's is about a quarter of the size of Founders. Tyler said:

> So, Vivant is the relatively small local, we are the mid-sized and Founders is the big leagues. But for our sustainable projects, the inspiration was city legislation. Then we wanted to be proactive and do it better than they could.

Short's desires sustainable growth, and they benchmark for energy, water, solar, and recycling. They have had some success in securing small grants and tax breaks for the sustainability work they do. Short's has a robust solar panel project in place, and they have reduced their food waste via a composting program. By doing dumpster checks, they added to the paper, cardboard, shrink-wrap, and wood pallets that they already recycle well. Their dumpster checks led them to realize that their number one waste was the disposable plastic gloves they use to examine beer and manipulate it during the brewing process. Tyler said:

> The gloves were probably the worst thing in our trash taking up the most space. We found a program that Kimberly Clark has to right cycle the gloves. Now we are a model, an example of what to do with that waste for other breweries, who also had a similar issue with the gloves. As a result, the dumpster is barely half full when they come to pick it up.

Short's has no large local livestock operations to sell their spent grain to, unlike some other Michigan breweries, but local farmers pay $5 for dump truck loads of it and the local vineyards take it for compost purposes. Short's uses their high-strength effluent in an off-site digester to make methane into electricity and an on-site treatment system to clean the rest of the equipment. Tyler wanted to make sure that I knew that Bell's (another West Michigan Brewery) does this at their brewery as well. This was an interesting insight about the sharing and collaboration of brewery sustainability projects between breweries. Tyler facilitates Short's sustainability committee, similar to the type that Brett and Liz hope to develop at Founders. Tyler got involved in this work because Short's was sending too much wastewater to the municipal treatment plant. Short's was pressured by the local government to do something about it. Tyler said:

> They were charging us and were going to charge us more. I did the math and we put a system to clean the wastewater at the end of the brewing process. It treats the low strength waste. Bell's system treats the high strength stuff at the beginning of the process.

Embarking on this wastewater project in late 2013, Short's had a lot of back and forth with the village. Short's water treatment project caused a real shake up in the village government because Short's was so successful. The Environmental Protection Agency wants industrial sites to pretreat their water, but the village wanted Short's to become a revenue source for the new water treatment facility they and the consultants were helping to create. According to Tyler, the new municipal plant is too big because it required planning for 30 years of possible growth overbuild. It cost the village $30–40 million, whereas Short's cleaned up its water by building a facility for $1.7 million.

Short's tried to do two things: first, they wanted to reduce their impact on the environment and second, they wanted to become more profitable in the process. Their costs are down and the pressure to be more sustainable now comes from Short's and Tyler's self-motivation as opposed to external pressure. When they got their wastewater treatment system up and running, they saw a 70% reduction in their wastewater production. Tyler says:

> Short's water sustainability projects were paid for by utilizing the minimal amount of grants and tax breaks there are in the state, locally and federally. No matter how good or cool these things are they always cost some money, but now our costs are down 35% since changing. We actually send cleaner water to the village, than the village puts in the lake. Short's is the dilution that is the solution to this area's water pollution.

Still, Short's echoed what others told me in that they want to make a high-quality product. That is what is first. Sustainability is second, and they do not really want that as a part of their marketing. It is about the context of the market a brewery is in. Tyler explained:

> For example, Sierra Nevada is practically no waste but you don't see that very easily in their marketing. It's the beer first, but it is also about the regulatory context the brewery is in as well. Ann Arbor, Michigan's up-to-date wastewater treatment facility means Jolly Pumpkin won't have to have what Short's and Bell's has. Lagunitas in California has one of the biggest brewery water treatment plants in the U.S. because that state wants you to do it—clean up your own mess. The Lagunitas in Chicago won't have its own water treatment because Chicago and Illinois want to do it, so they can charge to do it for you, because it makes the local municipality money—your wastewater is a revenue source.

Many of my interviewees implied or directly stated that the energy laws in Michigan are anti-net-metering, meaning that if you produce more energy than you take, the utilities would pay you. Many of my interviewees conjecture that this is because the big utilities, Consumers' Energy and DTE Energy, have a power monopoly and they do not want competition. Another barrier Tyler identified is that people always get mad when they do something new. Short's received some negative feedback from the surrounding community because some felt that grants should not exist and neither should solar panels. There is a contingent of beer consumers who say that because of the grants and tax breaks, they will not drink Short's beer ever again. But Short's and other breweries are taking significant strides toward sustainability and zero waste. This may be the crux of the problem of sustainability and craft beer tourism. They need to be able to attract, or at least retain, various types who may be repelled by sustainability projects, while telling everyone else about their innovations.

Brewery Vivant Interview Data

Brewery Vivant's website defines sustainability as:

> a word that gets talked about a lot, and in our opinion often times gets misrepresented. Instead of debating what the true meaning of it is, we thought we would just share what it means to us. Being a truly sustainable company means that we consider the impact of our decisions on the natural environment, the people that may be affected, and the financial health of our business. We want to balance all of these areas to operate our business with a long-term approach.

As a part of their effort to "beer the change they want to see in the world", and as a part of their Certified B corporation status, they have been offering a sustainability report for the past five years. Here are some highlights of their most recent goals in their posted report for 2015: 90% of their purchases are from within 250 miles of the brewery, 75% are from within Michigan, 50% of food is from within 250 miles, and 10% of the food is from Vivant farm. They are making strides to become zero waste to landfill and to get their water-to-beer ratio from their current 10:1 down to 3:1. Ten percent of their energy is on-site renewable although most of their reduction in carbon footprint to sales comes in the form of carbon offsets.

Kris, the owner and Director of Sustainability at Brewery Vivant, has a personal passion for sustainability, which is why the business uses the Certified B corporation model. Kris defines sustainability for the brewery as, "balancing the needs of people, planet and profit in decision making as well as being an active and supportive member of our local community". Kris acknowledged their water usage is something they need to get better at. To this point, they have allowed the brewers a lot of flexibility in how they do their work. But because they monitor these metrics, they are planning on introducing new equipment and policies to help reduce their water use. Kris said:

> As for pollution, I don't believe we are a polluting industry. The city monitors our effluent to ensure that it isn't negatively impacting their wastewater treatment process and so far they haven't had a concern with it. That said,

we are growing and are looking to divert more of our higher strength effluent from their system within the next year since we believe it will put us over their desired limits. Additionally, all of the storm water runoff from our property is treated on site through a septic-like system at the low point in our parking lot. This was a requirement from the city that was pretty cost prohibitive in order to create better water quality for all residents and avoid sewage overflows during heavy rains.

Sustainability is at the core of Vivant's business plan as a benefit corporation, and because they opened the business from scratch with these values, they have built a solid culture around sustainability with many employees seeking them out because of it. Still, other breweries have more resources to innovate on the water treatment issue. Kris says:

> As a small brewer, it is relatively hard to find solutions around energy and water efficiency that are affordable and fit within our limited physical footprint. There is a lot of opportunity to close the loops for small brewers, but most solutions have been targeted to large breweries. As the number of small breweries opening increases, we are starting to see companies look at this opportunity and develop solutions that are geared towards us, but that is just beginning.

Brewery Vivant works with the Michigan Brewers Guild sustainability subcommittee and many of the breweries in this area are trying to create the context in which the seemingly quirky sustainability innovations that work for each brewery become standard for the industry. Kris works hard to advocate for, and sustain, multiple community partnerships and campaigns:

> I also believe strongly in the power of the business voice in policy changes. We endorse various national campaigns like Brewers for Clean Water through the Natural Resources Defense Council and the Brewers Climate Declaration through Ceres. The better our water is for our business, the better it is for society. Brewers have a loud voice now in the media and by being vocal as a B Corp, we can use that attention to get these conversations out into the general public.

Vivant's social and community sustainability impacts are limited. Pay equity would seem to be a necessary component to sustainability. In a society where CEOs make so much more than their average workers, perhaps Vivant could address this issue by creating a policy ensuring that the executive compensation differential be no more than five times the average worker. Still, Brewery Vivant is "beering the change" they wish to see in the world and maybe their B corporation-oriented sustainability will spread too.

Discussion

Municipalities approached Founders and Short's because they were sending too much polluted water to the municipal treatment facility. But at this point, the breweries themselves are driving the change toward sustainability, creating a sustainable craft brewing industry. The brewers are self-motivated toward this end, although some interviewees acknowledged more squarely than the others the reasons behind these sustainability projects. In terms of water, the Clean Water Act and the Environmental Protections Agency require that water be pretreated by industry if water is being used by industry before they return it to the municipality. While many craft breweries come across as environmentally aware, there still needs to be a regulatory framework that acts as the stick to the sustainable carrots provided by voluntary professional organizations and alternative business models, like the B corporation. Confirming some of the corporate social responsibility literature (Chouinard & Staley, 2012), Brewery Vivant is aware of their water use because they want to "beer the change". In response to the first draft of this study, Vivant communicated that the Michigan Department of Environmental Quality is getting much more proactive in helping businesses identify ways to improve. Founders and Shorts were prodded to create sustainable water systems, and after the fact, Short's delivers cleaner water than they receive from the municipal water supply. In this instance, the governmental regulatory regime on industry pushed it in a sustainable direction. When prodded, these breweries desired to exceed the environmental standards set by the current regulatory regime, despite the lack of governmental incentives.

In many ways, this study confirms the questionable prospect that brewing, sustainability, and tourism are more than marketing (Mowfurth & Munt, 2016). While all three of these breweries are "destination" breweries for beer tourists and others, Founders and Short's lead with their beer. They do not hide their sustainability efforts. They also do not prominently display and market their sustainability projects. The crux of the problem of sustainability and craft beer tourism is the need to be able to attract, or at least retain, a diversity of customers, some of whom are automatically repelled by sustainability.

These craft breweries contribute to two aspects of the triple bottom line of sustainability—environment and economy. Confirming literature regarding environmental justice and sustainability, these breweries are advancing self-interested, cost-saving ways to reduce natural resource use to better the environment and save money for their business economies and are not moving the needle very far for people, equity, or social justice (Agyeman et al., 2003; Kolbert, 2014; Whitehead, 2007). By leading with producing high-quality beer, by using high-quality sustainable resources and processes, breweries address the environmental and economic components of sustainability. It becomes much less clear what they do to address the social justice and social equity components of sustainability, especially in their attempts to be as apolitical as possible when promoting sustainability.

Conclusion

In these challenging times, it is important to investigate how businesses, consumers, and citizens take a complex value-laden project, like sustainability, and attempt to implement it—to take a theory and make it reality. Change to business as usual is necessary at multiple levels. The willingness of these breweries to make the changes they can and then work to institutionalize them is extremely important to observe. These three cases give important insights on how to start work on complex environmental, economic, and social problems. This study creates the framework for future research that should compare more breweries, the Michigan Brewers Guild, the Brewery Association, the Sustainability Subcommittee, the Michigan Department of Environmental Quality, and related private, public, and non-profit organizations in Michigan and elsewhere to create a local, regional, national, and global study to understand how breweries in other regions operate, compare, and "do" sustainability.

References

Agyeman, J., Bullard, R., & Evans, B. (2003). *Just sustainabilities: Development in an unequal world*. Cambridge: MIT Press.

Alworth, J. (2015). *The beer bible*. New York: Workman Publishing.

Averill, G. (2016 January/February). Brewers are our most creative conservationists. *Outside Magazine*, p. 74.

Brown, P. (2003). *Man walks into a pub: A sociable history of beer*. London: Pan Macmillan Books.

Brown, P. (2006). *Three sheets to the wind: One man's quest for the meaning of beer*. London: Pan Macmillan Books.

Brown, P. (2009). *Hops and glory: One man's search for the beer that built the British Empire*. London: Pan Macmillan Books.

Canham-Nelson, M. (2012). *Teachings from the tap: Life lessons from our year in beer.* Caramel Valley, CA: Beer Trekker Press.

Chouinard, Y., & Staley, V. (2012). *The responsible company: What we've learned from Patagonia's first 40 years.* Ventura, CA: Patagonia Books.

Jackson, M., & Lucas, S. (2000). *Michael Jackson's great beer guide.* London: Dorling Kindersley.

Kolbert, E. (2014). *The sixth extinction: An unnatural history.* New York: Picador.

Kouwenhoven, J. (1961). *The beer can by the highway: Essays on what's "American" about America.* Baltimore: Johns Hopkins University Press.

List of Countries by beer consumption per capita. (n.d.). In *Wikipedia*. Retrieved October 15, 2016, from https://en.wikipedia.org/wiki/List_of_countries_by_beer_consumption_per_capita

Lorr, M. (2012a). *The popularization of sustainable urban development: Chicago, Vancouver, and marketing environmental and spatial justice in an era of neoliberalism.* Unpublished Ph.D. dissertation, University of Wisconsin-Milwaukee Urban Studies Program.

Lorr, M. (2012b). Defining sustainability in the context of North American cities. *Nature & Culture, 7*(1), 16–30.

Mowforth, M., & Munt, I. (2016). *Tourism and sustainability: Development, globalisation and new tourism in the third world.* London: Routledge.

Neu, D. (2011). *Chicago by the pint: A craft history of the windy city.* Charleston, SC: The History Press.

Nodrum, A. (2015, March 17). Craft beer breaks double digit market share for the first time in US. *International Business Times.* Retrieved from http://www.ibtimes.com/craft-beer-breaks-double-digit-market-share-first-time-us-1849648

Oliver, G. (2012). *The Oxford companion to beer.* Oxford: Oxford University Press.

Pure Michigan. (2013, May 1). Three reasons why Grand Rapids is beer city USA. Retrieved from http://www.michigan.org/blog/guest-blogger/three-reasons-why-grand-rapids-is-beer-city-usa/

Pure Michigan. (2015, November 19). Become a "brewsader" Grand Rapids beer city passport. Retrieved from http://www.michigan.org/blog/region/become-a-brewsader-with-the-grand-rapids-beer-city-passport/

Rosenzweig, R. (1991). The rise of the saloon. In C. Mukerji & M. Schundson (Eds.), *Rethinking popular culture: Contemporary perspectives in cultural studies* (pp. 121–156). Berkeley: University of California Press.

Ruschmann, P., & Nasiatka, M. (2006). *Michigan breweries.* Mechanicsburg, PA: Stackpole Books.

Whitehead, M. (2007). *Spaces of sustainability: Geographical perspectives on the sustainable society.* London: Routledge.

CHAPTER 5

Representing Rurality: Cider Mills and Agritourism

Wynne Wright and Weston M. Eaton

INTRODUCTION

Representations of the rural evoke a plethora of images, but for many, positive imagery of simplicity, dense social ties, and bucolic landscapes come to mind. Whether Americans envision rugged settlers taming the prairie, the thriving community of Main Street, or the face of God in nature, the rural idyll has been a staple in American consciousness. This *idyllization* of rural spaces and livelihoods can also be accompanied by a reverence for, and an animation of, many rural customs and artisan practices that inspire collective admiration and conjure up notions of authenticity.

A previous version of this chapter was presented at the International Rural Sociology Association (IRSA) meeting in Toronto, CA, August 2016. Funding for this research was supported by Michigan AgBio Research.

W. Wright (✉)
Departments of Community Sustainability and Sociology,
Michigan State University, East Lansing, MI, USA

W.M. Eaton
Department of Agricultural Economics, Sociology, and Education,
The Pennsylvania State University, University Park, PA, USA

At the same time, obdurate reverence for, and enactment of, the rural idyll can mask modern-day realities. Little (1999) contends that, too often, the rural idyll has "served to detract from the recognition of variety and, indeed, alongside the concept of 'otherness', to simplify our understanding of power relations within rural society and of the contestation of the reality and representation of rural culture" (p. 440). Hinrichs (1996) adds that idealized rural images evoke tradition in ways that omit tension, diversity, and complexity; "notions of rural tradition dwell selectively on its most sanitized, beneficent possible features" (p. 263). The coexistence of such contradictory tensions provides an opportunity to explore the rural idyll further.

In this chapter, we explore the role that apple cider mills, organized as farm tourism destinations, play as agents of rural cultural representation. We probe the degree to which these vibrant agritourism destinations can be enrolled to serve as a symbolic vehicle of contemporary agriculture and rurality. Cloke (1997) writes, "Many people are likely to 'know' rural areas more through watching popular television programs than through personal experience" (p. 372). If accurate, a visit to an apple cider mill may be one of the few opportunities the public has to experience agriculture beyond the realm of eating. Therefore, venues such as cider mills can be viewed as important arenas for understanding how agriculture and rural life are constructed and performed for uninitiated, yet politically salient, and economically important audiences.

Like all farm tourism, apple cider mills package, accentuate, and commoditize the social and cultural value in apple farming. Agritourism has grown in recent years, rising 6 percent annually in North America and Europe from 2002 to 2004 (Choo, 2012). Advocates argue that it brings "fun" to the farm (George & Rilla, 2011), yet most see its importance as an economic necessity alleviating constraints placed on family farms by the productionist agro-food model (Che, Veeck, & Veeck, 2005). Research has found agritourism to be a stimulus to farm family's financial stress and risk management (McGehee, Kim, & Jennings, 2007), rural development (Hinrichs, 1996), nature conservation (Lane, 1994), and cultural consumption (Che et al., 2005). Moreover, it is rooted in a contemporary theoretical turn that privileges rural development processes valorizing local resources and rejuvenating human and ecosystems (Ploeg et al., 2000). Recent literature suggests that agritourism not only fosters economic development, it can also contribute to the maintenance and reinforcement of the rural social fabric, as well as the preservation of the environment.

Given this promise, we see the nexus of symbolic representation and apple cider mill tourism as fertile terrain for an exploratory investigation.

We begin by examining the literature on rural representations in order to query agritourism as a symbolic vehicle of agriculture and rurality (Halfacree, 2007). Representation, or "mental perception of the countryside", is often central to rural tourism as tourists reactivate "well-established stereotypes about nature and purity" firmly embedded in their "collective consciousness" (Bessière, 1998, p. 20). Our concern is with the ability of cider mills to shape meaning and understanding of agriculture and rural life for tourists drawn from a generation whose knowledge of these domains is limited or comes through the realm of consumption/entertainment rather than production or livelihood. Following Murdoch and Pratt (1993), we see apple cider mill agritourism entrepreneurs as "actors [who] impose 'their' rurality on others" by choreographing, staging, and performing educational and leisure farm activities (p. 411).

This platform to construct rurality and commodify rural culture for tourists raises important questions that may challenge popular understandings of what constitutes a sustainable form of rural life. The chance to represent agriculture and rural life to tourists and consumers may provide producers a renewed opportunity to tell their story—to re-emerge as authoritative spokespersons of rural life—roles currently held by outsiders, such as policy makers or actors further up the supply chain. Is it possible that cider mill agritourism might permit a new form of power to which the producing class has increasingly struggled to access given the rise of the consumer? We conclude this chapter by discussing some of the implications of current-day apple cider mill representations.

THEORETICAL OVERVIEW

The return to the study of the rural over the past two decades is accompanied by a rising interest in neolocalism. Shortridge (1996) refers to neolocalism as an exercise in forging "geographical identities" (p. 10), an effort to establish dense social ties and identities connected to particular places. Flack (1997), too, argues that desire for a sense of place is at the heart of this localism turn. He offers up the rise in microbreweries as evidence of "American's rootless angst" or "self-conscious reassertion of the distinctively local" (p. 38). But why have so many Americans become concerned about the local?

Holtkamp, Shelton, Daly, Hiner, and Hagelman (2016) frame the neolocal movement as a form of popular resistance to the "homogenization of the economy and urban landscape" (p. 66). Others go further to contend that the rise in the neolocal movement is a response to frustration with issues of scale—in an effort to counter globalization processes that are believed to obscure transparency, dilute distinctiveness, and threaten authenticity. Thus, for some, it is viewed as a repository of perceived simplicity and a respite from the so-called complicated and fast-paced modern world where the standardizing forces of neoliberalism erode social relations and undermine commitments to place and cultural attachments. In short, it is a retreat from the complexities, uncertainties, and vulnerabilities of modern life.

It is important to draw the parallel between the resurgence of the neolocal movement and the return to the rural as we see these two forces as interconnected (Ploeg et al., 2000), yet not synonymous. While much of the localism movement, especially as it pertains to the demand for local food and drink, is rooted in the rural, any reverence for the countryside is often eclipsed by efforts to draw attention to the urban, urban farming is just one example. This explains why we have not seen similar robust efforts to eradicate rural poverty, to take seriously the challenges faced by rural educational systems, to counter the ongoing decline of farm families, or to address the dearth of resources made available to rural communities for planning and development. Much of the neolocal research has been focused on the popularity of microbreweries, and while certainly a commodity with agricultural roots, much of the analysis has tended to emphasize the rise of consumer demand and the geographical location of microbreweries in urban and suburban landscapes. We acknowledge that apple cider mill tourism has benefitted from the neolocal movement, but such farm tourism venues significantly pre-date the post-1980s neolocal movement. Moreover, apple cider mills, unlike other neolocal revival projects, are uniquely rural in their constitution. As such, theories of rural representation hold more utility for understanding the social construction of apple cider mills.

Theories of social representation of the rural have become a growth industry over the past three decades (Cloke, 1997). The deconstructive turn advanced by post-modernism sparked renewed interest in the rural through attention to the social construction processes that make it possible (Halfacree, 1993). The intellectual turn to culture and agency via phenomenology and the sociology of knowledge (Cloke, 1997), extended to the rural, accentuate the process by which people creatively shape reality through everyday interaction and imaginaries (Halfacree, 2007). From this intellectual tradition, rurality arises from "the social production of a set of meanings" attributed to rural spaces, peoples, and practices (Mormont, 1990, p. 36).

Foregrounding rural social interaction over spatial or materialist dimensions sets the stage for understanding rurality as a dynamic "social construct and 'rural' becomes a world of social, moral, and cultural values in which rural dwellers participate" (Cloke & Milbourne, 1992, p. 360). This approach to the study of rurality has allowed scholars to probe "how practice, behavior, decision-making and performance are contextualized and influenced by the social and cultural meanings attached to rural places" (Cloke, 2006, p. 21), thereby expanding our capacity to understand the realities of rural people.

Such work foregrounds the microelements of social life, such as language, symbols, and social norms, the rural as imaginary or an ideal, and the situatedness of everyday experience (Cloke, 2006). Everyday words, symbols, and actions become tools in a socialized arsenal to make meaning and represent rural selves to others. Halfacree (1993), for example, argues that the rural is best represented through discourse—through the "words and concepts understood and used by people in everyday talk" (p. 29). Through discourse, it becomes evident that meanings of rurality do not inhere in the material but are socio-psychological constructs (Cloke & Milbourne, 1992). Edensor (2006) centralizes the role of action in rural representation with the performance metaphor where rural dwellers "perform" rurality—or behaviorally manage an impression of themselves as rural people—with their bodies, discursive practices, material artifacts, and social environments. In short, spaces become a theater where actors don costumes, stage the setting, and enact performances with culturally appropriate props and scripts and in doing so, rurality is constructed. In the tourism context, the goal is to "produce affective, sensual and mediatized experience—within a format of 'edutainment'" (Edensor, 2006, p. 488).

Background

The food and agriculture sector is central to Michigan's economy. It generates $91.4 billion in economic activity annually and employs 923,000 people or 22 percent of the labor force (Knudson & Peterson, 2012). Surrounded by four of the five Great Lakes, much of the state enjoys a unique micro-climate that is acutely suited to the production of fruits and vegetables, themselves highly amenable to agritourism. Michigan is the third largest apple producing state in the nation, and apples are the primary fruit grown, producing 24 million bushels each year. Most Michiganders come to know apples either through their local supermarket or via annual fall family outings to a local orchard and cider mill where apples, donuts, and cider are common fare. According to the Michigan Apple Committee (2016), there are 135 apple orchards/cider mills in the state.

The state's unique geographical location has also enabled a vibrant tourism economy that capitalizes on markets created by travelers visiting the state for lakeside holidays. As a result, apple cider mills have long been a fixture in Michigan. The marriage of fruit production with tourism helps explain, in part, why most of the cider mills are disproportionately located along the Lake Michigan coastline. "Michigan farmers are utilizing agritourism as a value-added way to capitalize on their comparative advantages, their diverse agricultural products, and their locations near large, urban, tourist-generating areas" (Che et al., 2005, p. 225). The Michigan Department of Agriculture and Rural Development offers a printed and online agritourism guide to these venues to help tourists locate fresh apples and other produce. The Select Michigan campaign also helps consumers identify Michigan-grown foods by providing a label to be affixed to locally produced and processed products.

Traditionally, the classic cider mill tourism model welcomed visitors to the farm primarily for sales of apples and cider. In recent years, however, selling such farm staples has become insufficient. Consumers are increasingly provided an experience, including education and entertainment, in addition to the chance to purchase an array of farm products. Today's cider mill proprietors are installing u-pick apple orchards, farm markets, pumpkin patches, offering wagon rides, petting zoos, restaurants, pie barns, children's playgrounds, novelty shops, and more. Others transform their farms into wineries, cideries, distilleries, and tasting rooms, intended to attract new groups of consumers with increasingly epicurean tastes. Regardless of the diversification strategy, the apple remains the center of branding and farm marketing even though the on-farm production of apples is declining.

This chapter now turns to consider how apple cider mills represent agriculture and rural life. We contend that cider mills are sites where actors choreograph performances and stage settings meant to communicate specific imagery and meanings and, in this way, influence—directly or indirectly—public perception. The following section considers the exploratory methods employed in this study.

Research Methods

This study is an instance of exploratory, or grounded theory, research in which our objective was to discover new insights about how cider mills appropriate images, stage artifacts, and engage in performances associated with rurality. Goulding (2002) advocates for the adoption of "grounded theory when the topic of interest has been relatively ignored in the literature or only have been given superficial attention" (p. 55). We contend that the lack of scholarship on cider mills as agents of rurality fits this criterion. For these reasons, we adopt a broad lens in order that salient issues and/or variables might be identified.

The research on which this chapter is based was undertaken from 2013 to 2014 as part of the lead author's broader study on value-added agriculture as a rural development strategy. The data on which this chapter is built draws upon interviews with Michigan cider mill owners, observation, and document analysis. One-on-one interviews were conducted with seven cider mill owners/managers at the site of the cider mill and ranged in length from one to two hours. Interviews consisted of approximately 40 open- and closed-ended questions covering subjects such as farm history, farm/mill changes, management strategies, visions, and relations with consumers. We also made participatory observations of Michigan apple cider mills. Each author assumed the role of a tourist and took part in a minimum of five cider mill visits, along with other guests. Finally, documents included for analysis consisted of the cider mill's website, signage, brochures, as well as product labels.

Representing Rurality

In this section, we examine efforts to organize—or choreograph—the interaction between the cider mill and guests, as well as elements of staging and performance that are used to animate rural life and agriculture. We consider instances of how cider mills reproduce nostalgic imagery associated with an agrarian past and an escape from modernity, along with perceived traditional social relations that accompany such imagery.

Cider Mill Customs

Understanding agriculture and rural representation by cider mills appears to begin before one steps foot on the farm. A cursory glance at the branding and marketing of these destinations suggests that communicating tradition and custom is central to cider mill identity. For instance, one of the earliest encounters tourists have with a cider mill is in the farm name. Farm names speak to us and provide a contextual hint, foretelling their historical relations, landscape characteristics, or the identity of those populating the farm.

One glance at the directory and it quickly becomes clear that the vast majority of orchards in Michigan follow an old custom of naming the business for the family of owners. Blake's Apple Orchard, Miller's, Parker's, Yates', Anderson's, Friske's, and the list of family surnames bestowed on the mill continues. In these cases, cider mill tourism takes on a time-honored custom by attaching itself to family relations and family-farm agriculture. Cider mill tourism in this post-agrarian era is also constructed as a human or family endeavor eclipsing the role of nature or any community effort, as well as masking highly politicized industrial aspects of agriculture. Moreover, the use of family names serves to masculinize the cider mill, elevating male heads of households over women entrepreneurs. The adoption of family surnames communicates norms and values that undervalue women's identities and the contributions they make to community sustainability. Over time, such masculine naming strategies come to be seen as benign, even invisible. Carrington, Donnermeyer, and DeKeseredy (2014) argue that one of the prevailing representations of the rural, in general, is that the "dominance of man and mankind over women and nature, is represented as natural, and unproblematic" (p. 467).

In addition to identifying the mill by the owner's surname, a second nod to customs frequently seen in the name is to draw upon informal or folksy monikers. "Uncle John's Cider Mill", "Grandpa's Cider Mill", "The Country Orchard", or "Apple Charlie's Orchard and Cider Mill" illustrate this practice. Others, like "Robinettes Apple Haus and Winery", use non-conventional spellings that have more to do with evoking a sense of nostalgia for a bygone era rather than with the owner's ethnic heritage. These monikers rest on stereotypes of white, rural clichés that "other" such venues as simple, folksy, or a throwback to another time. The image of an intoxicated country "bumpkin", dressed in overalls—the stereotypical rural uniform—is such an instance (see Fig. 5.1). Such representations lend themselves to sanitized consumer images, whereby people are mere

Fig. 5.1 Uncle John's Cider Mill billboard image

caricatures in a rural spectacle. These labels blur the reality of cider mill operations, casting a shadow on any attempt to understand the modern complexities of the agricultural industry; it can remain a space of joy or entertainment for the consumer where cider mills proprietors and their employees exist in an idealized space to entertain and fulfill the urban desire for bucolic nostalgia.

If the business name has set the stage for an idealized representation of cider mills, this practice is reinforced in marketing materials that accentuate heritage. Brochures and online marketing media begin the process of welcoming the guest to a family-owned cider mill that has long roots. This nod to the durability of the cider mill coincides with a broader ideology that rural communities are signifiers of stability. For example, Dexter Cider Mill proudly boasts that it "is the oldest cider mill still operating in the entire state". Others tout their history of having been "established in 1895" (Franklin Cider Mill) or that they are a "fifth-generation farm" (Uncle Johns) or are "celebrating their 89th year" (Erwin Orchards). Tradition and longevity are enrolled, in part, to establish boundary markers, demarcating insiders from outsiders. In rural locales, embeddedness in tradition, and long-standing community ties, signals insider status. This boundary reaffirms old timers as authentic sources of rural knowledge and practices, while maintaining social cohesion and subverting the influence of those less well established. While advertising materials enroll tradition and custom frequently to represent themselves, they also construct cider mills as distinctive, setting them apart from their competition.

With a history that dates back to 1863, Yates Grist Mill opened its doors beside the rapidly flowing waters of the "then" Clinton-Kalamazoo

Canal. Amid the beautiful countryside of Rochester, Michigan, the memories began in full bloom. By 1876, the Yates family installed a cider press into the existing water-powered process and began producing delicious Michigan cider... Local farmers, orchard owners, and landowners would bring their apples to Yates for custom apple pressing. Over all these years, Yates has been producing the same kind of fresh 100% all natural cider that folks enjoyed way back in 1876... Good fruit makes a difference and you will see and taste the difference at Yates. Will it always taste the same? Nope... As the season progresses, more varieties ripen. That's what makes the cider season interesting. As each week passes, different cider flavors can be blended to round-out the taste (Yates Grist & Cider Mill, 2016).

Blake Farms is a family-owned and family-operated orchard and cider mill rooted in strong family traditions and a commitment to excellence. Our historic farm started in 1946 by Gerald and Elisabeth Blake and their 13 children. It was one of the very first "Pick Your Own" orchards in Michigan. Blake's are famous for their award-winning apple cider and have been voted the number one apple orchard in Michigan by American Automobile Association authorities (Blake's Apple Orchard, 2016).

In the above references, tradition is framed through the familiar embrace of family, durability, hard work, and community embeddedness to produce high-quality cider. Such framing lowers the boundaries of formality by evoking a down-home character associated with the work ethic ascribed to rural people. They also accentuate social solidarity and the expressive forms of rationality associated with rurality, not the instrumentality of science and industrial cider production methods, common techniques of production reserved for the backstage.

At the same time, these representations illuminate distinctive attributes of cider mills by highlighting production practices as well as situating them within the natural landscape, furthered by a nod to seasonality. Imagery such as "flowing waters", "beautiful countryside", "cider season", "the first Pick Your Own orchard", and the "Clinton-Kalamazoo Canal" also help to brand this orchard as uniquely embedded in place. Both production and geographic distinctiveness is punctuated throughout and further the work of segregating orchardists from tourists and, in this way, perpetuating their image as unique, or "other" (Little, 1999). For Weightman (1987), "the tour brochure directs expectations, influences perceptions, and thereby provides a preconceived landscape for the tourist to discover" (p. 230). This suggests that what tourists may be primed to witness is a cultural reproduction of the simplified rural/urban binary at work. Orchardists may foreground that which differentiates rural and urban dwellers (e.g., nature, culture, heritage, etc.) instead of what binds them (e.g., modernity, neoliberalism, etc.). Therefore, by emphasizing differences between rural and urban dwellers, proprietors may be underscoring tourists' preconceived ideas of rural life and reinforcing rural/urban cleavages where common values, identities, and aspirations are otherwise masked.

Consuming the Rural

The advertising brochure and the website illuminate tradition and distinctiveness, but when guests arrive they enter a scene designed and staged to reinforce this imagery. From the architecture tourists observe, to the artifacts adorning the landscape and buildings, as well as the products on offer for sale, tradition and cultural heritage are enrolled to represent life on a working apple orchard and cider mill.

Perhaps no image resurrects rurality in the same way as agricultural barns constructed of materials typical of the nineteenth century, harvested from the local terrain, with massive hand-cut tongue and groove wooden beams. Barns, as a form of heritage capital, are typically centerpieces in Michigan apple cider mill tourism. Increasingly, the apple orchard itself tends to be less visible as more and more growers elect to purchase a significant share of their produce from off the farm, thereby making the orchard itself more of an aesthetic fulcrum than a critical productionist centerpiece. In its place, the barn—typically the site of consumption—has taken center stage. Modern pole barns built of enameled metal and chemically treated timbers do not conjure up the same rural aesthetic as do grand, century-old barns erected in the architectural tradition of the community's European ancestors.

From a former site of production, old barns are today repurposed as spaces for urban consumption (Hinrichs, 1996). Consider, for instance, the ubiquity of jars of jam, maple syrup and pickles, woven rugs, and other country-style goods for sale at cider mills—but also the growing trend toward incorporating value-added fruit products in the form of wine, hard cider, distillates, and other fermented beverages. Spurred by the thriving US craft beer industry, the popularity of fermented products like hard cider is booming nationally, and especially in fruit-producing states like Michigan. As is increasingly evident, adding craft beverages like hard cider to a consumptive repertoire can ease cider mills' transition from traditional sites of agricultural production into consumptive spaces. Moreover, in reorganizing barns as cideries and wineries—spaces for production and consumption—cider mills and their barns, orchards, and farms are again transforming, now into hybrid consumption/production sites.

Consider Robinette's Apple Haus & Winery and Uncle John's Cider Mill, two leaders in the revival of Michigan's hard cider industry, both of which recently converted farm structures into such hybrid spaces (in 2006 and 2003, respectively). For instance, Uncle John's, which initially built its barn or "Fruit House" in 1918 to store and package fruit for wholesale and retail, and was used as a workshop in the interim, is today reborn as a dual space. A cidery and distillery has been built and sectioned off to house production, packaging, and storage, while in an adjacent but separate "tasting room", employees trained in showcasing the cider mills' fermented products serve samples and sell bottled products.

Tasting rooms extend the on-the-farm consumption theme of cider mills' ancillary bakeries, fruit markets, and country-style goods. Here, customers peruse and sample wines, ciders, or distillates in settings staged to remind tourists of the product's imagined agricultural roots. Consider, for instance, the use of wooden wagons or oak barrels as product displays or antique apple presses staged alongside empty fruit crates.

Bottle labels serve a similar performative purpose. Labels tend to showcase iconic images of farm life—blossoms on tall, expansive fruit trees, or farmers mounting wooden ladders to harvest perfectly ripened fruit. Staging such images on package labels gives the impression that the essence of farm or orchard life can be found within the bottle, and that one can experience this lifestyle via the act of consumption. The "Chapman's Blend" label from Vander Mill provides a striking example. Here, we find Johnny Chapman—aka Johnny Appleseed—striding across the label with shovel and seed bag at hand, a loaded apple branch in the foreground reminding us of his legacy in American literature. As food writer Michael Pollan (2001) suggests, the Chapman image "has the resonance of myth":

> *It's the story of how pioneers like him help domesticate the frontier by seeding it with Old World plants. "Exotics," we're apt to call these species today, yet without them the American wilderness might never have become a home.* (p. 5)

This mythical ideal is then brought full circle with the tagline "An American Heirloom Cider" (Fig. 5.2).

Fig. 5.2 Vander Mill Chapman's Blend label

The beverages behind labels and inside bottles vary widely in terms of both ingredients and production practices. For instance, cider mill wineries and cideries may produce their fermented products on-site, such as in the converted barns discussed above. It is not an uncommon practice, however, for cider mills to contract with others to produce wine or cider from the mill's fruit, to rely upon or incorporate juice purchased from growers across the region, or for products sold under a cider mill's label to be produced off-site altogether. However, regardless of actual material production practices, all cider makers tend to appeal to rural and agricultural imaginaries and icons when marketing their products.

However, a closer look at the reality of craft beverage production practices reveals an inherent tension between rural idylls and the technological requirements of producing a standardized commodity like wine or hard cider. For instance, while the naturally occurring yeasts found in unpasteurized cider will indeed covert the sugars in the juice into alcohol, professional cider makers rely, to various degrees, on industrial techniques and ingredients (e.g., sanitization and yeasts produced in laboratories) to produce standardized or, at a minimum, controlled products capable of maintaining intended flavor and aroma profiles through the rigors of storage, delivery, and retail display.

These realities of modern production processes present both challenges and opportunities for cider mills and makers who employ modern techniques while appealing to rural idylls in their marketing approaches. The bucolic imagery diversion of tasting room and label choreography, which diverts consumers' gaze from modern practices, is one way this tension is navigated. Other cider mills and producers instead retune modern production practices to directly incorporate pre-industrial fermentation practices, including fermenting and aging cider in oak barrels.

Oak barrels symbolize "old ways" of making cider, when farmers would press apple harvests directly into barrels, perhaps those that had held whiskey or bourbon, add maple syrup, honey, or other locally available sugars, allow the naturally occurring wild yeasts to ferment the sugars into alcohol, and store the product (termed "scrumpy") in the barn or cellar through the fall and winter. Drafts would then be taken directly from the barrel's wooden spigot. Like fermenting cabbage into sauerkraut, or dry curing meats, the success of scrumpies depended on the presence (and absence) of bacteria and other factors, like temperature and humidity, not entirely within the control of the farmer, meaning batches would vary widely from year to year, farm to farm.

Today's hard cider makers who undergo the investment of time and resources to incorporate a barrel program cherish this unpredictability—framing their inevitable few failed batches as necessary tolls on the path of authentic craft production—as do their epicurean consumers who prize unique products. Arguably, and as a testament to the symbolic power of rural idylls, products that pass the test of true authenticity with the growing ranks of hard cider enthusiasts do so not only through material production choices like incorporating oak barrels, but perhaps more importantly through successful symbolic appeals to myths of rural American farmers and their oak barrels, spontaneously fermented cider, and now bygone ways of life.

Discussion

Our argument above has accentuated the ways modern cider mills represent idealized forms of a bucolic lifestyle. The popularity of apple cider mills—and related agritourism venues—demonstrates the salience of the rural idyll for both producers and consumers. As Short (2006) puts it, "the consumer buys (into) the countryside through the link made with products, whether they be cars, duvets, beer" (p. 143) or, in our case, apple cider. But the popularity of the rural idyll in the twentieth-first century demands consideration of the implications of this reverence for rural mythology. Such representations have consequences that are important for cider mills, their employees as well as their consumers, but also more generally for the relationship between agricultural producers and citizen consumers.

Many of the instances provided above misrepresent modern rural life. Idyll rural representations skew authentic rural realities and, as such, limit opportunities for agricultural producers to engage urban citizens/consumers on pressing social, economic, and environmental matters. In doing so, this perpetuates misunderstandings as to the contemporary realities of agricultural production, including apple and cider production. Guests may leave the apple cider mill under the perception that rural dwellers live the "good-life", surrounded by the stability of family, and with the beauty of nature at arm's reach. Missing is any opportunity to convey the complexities of modern-day apple farming, or rural life more generally. Nowhere will they be able to learn that mill owners and their families engage in an 80-hour workweek just to keep their economic heads above water nor will anyone communicate to the guest information of the economic crisis that has faced these growers over the past two decades and the declining number among their ranks. Guests will also leave the cider mill without having been told that most of the workers who have been waiting on them during their visit that day work as seasonal employees for minimum wage and have no 401(k) or other retirement safety net. They will not have the opportunity to consider that the tranquility and peace of the rural community is quiet because all of the industry and shops that are vital to a thriving Main Street are closed due to a stagnating economy. It is doubtful they will appreciate that after a long day of work at creating a joyful and carefree experience for the guest, that a lucky few employees will drive 20–30 miles to the nearest regional center to work a second job. In such ways, the positive rural idyll of the apple cider mill can misinform as much as it can evoke tradition and nostalgia. By focusing on the bucolic, attention is diverted from meaningful and pressing rural challenges.

We might also examine pressing issues of apple harvesting equipment and the implications of technological changes specifically for apple producers. Namely, due to concentration in the industry, apple producers are under increasing pressure to increase yields and reduce costs, especially labor. These pressures have led to increased reliance on emerging technologies, such as disease-resistant apple varieties and harvesting equipment, both of which serve to increase efficiency. However, these technologies impose new forms of economic risk on apple producers.

These disparate examples begin to illustrate the contemporary complexities of agricultural production modern cider mills must navigate. And yet, these or other examples are rarely conveyed to urban consumers, who instead confront rural representations that serve to mask these contemporary challenges. Could this be otherwise? Perhaps growing consumer interest in organic, local, and "craft" products can be read as indication that consumers are interested not only in an idyll past, but also in a new future for food and beverage production? One premised on lifting the bucolic veil in order that consumers might witness and consider the inner workings of agricultural production so that they are better prepared to carry out their consumer and citizenship roles?

Conclusion

As we have shown, modern cider mills represent idealized forms of bucolic nostalgia. This chapter has shown that tradition, cultural heritage, and distinctiveness play key roles in cider mill representations of rurality. This is accomplished in the choreography and staging of the cider mill experience. At the same time, our findings support previous research that has found farm tourism to be "inextricably intertwined with historical, political, and cultural processes" (Pritchard & Morgan, 2001, p. 168). While engaged in the representation of idealized rurality, missed opportunities abound to build authentic representations and showcase modern rural realities. The absence of such authentic realities is especially troubling for vulnerable populations, such as cider mill proprietors and their workforce who are often struggling to make a living wage. This study suggests that if cider mills are to continue to prove durable, inroads must be made into bridging lacunas, giving consumers/citizens the opportunity to engage in meaningful and authentic learning opportunities so that they can become a better-informed force for reshaping rural communities and the food system. We see this discrepancy between imagery and reality as a fruitful line for future research.

References

Bessière, J. (1998). Local development and heritage: Traditional food and cuisine as tourist attractions in rural areas. *Sociologia Ruralis, 38*(1), 19–34.

Blake Farms. (2016). Retrieved November 15, 2016, from http://blakefarms.com/blakes-orchard-and-cidermill

Carrington, K., Donnermeyer, J. F., & DeKeseredy, W. S. (2014). Intersectionality, rural criminology, and re-imaging the boundaries of critical criminology. *Critical Criminology, 22*(4), 463–477.

Che, D., Veeck, G., & Veeck, A. (2005). Agriculture and the selling of local food products, farming, and rural America tradition to maintain family farming heritage. In S. Essex, A. W. Gilg, & R. B. Yarwood (Eds.), *Rural change and sustainability: Agriculture and environment and community* (pp. 109–121). Wallingford: CAB International.

Choo, H. (2012). Agritourism: Development and research. *Journal of Tourism Research and Hospitality, 1*(2), 1–2.

Cloke, P. (1997). Country backwater to virtual village?: Rural studies and 'the cultural turn. *Journal of Rural Studies, 13*, 367–375.

Cloke, P. (2006). Conceptualizing rurality. In P. Cloke, T. Marsden, & P. H. Mooney (Eds.), *The handbook of rural studies* (pp. 18–28). London: Sage.

Cloke, P., & Milbourne, P. (1992). Deprivation and lifestyle in rural Wales: Rurality and the cultural dimension. *Journal of Rural Studies, 8*(4), 71–76.

Edensor, T. (2006). Performing rurality. In P. Cloke, T. Marsden, & P. H. Mooney (Eds.), *The handbook of rural studies* (pp. 355–364). London: Sage.

Flack, W. (1997). American microbreweries and neolocalism: Ale-ing for a sense of place. *Journal of Cultural Geography, 16*(2), 37–53.

George, H., & Rilla, E. (2011). *Agritourism and nature tourism in California*. Davis, CA: University of California Press.

Goulding, C. (2002). *Grounded theory: A practical guide for management, business and market researchers*. London: Sage.

Halfacree, K. H. (1993). Locality and social representation: Space, discourse and alternative definitions of the rural. *Journal of Rural Studies, 9*(1), 23–37.

Halfacree, K. H. (2007). Trial by space for a 'radical rural': Introducing alternative localities, representations and lives. *Journal of Rural Studies, 23*, 125–141.

Hinrichs, C. C. (1996). Consuming images: Making and marketing Vermont as distinctive rural place. In P. Vandergeest & E. M. Du Puis (Eds.), *Creating the countryside: The politics of rural and environmental discourse* (pp. 259–278). Philadelphia, PA: Temple University Press.

Holtkamp, C., Shelton, T., Daly, G., Hiner, C. C., & Hagelman III, R. (2016). Assessing neolocalism in microbreweries. *Papers in Applied Geography, 2*(1), 66–78.

Knudson, W., & Peterson, C. (2012). *The economic impact of Michigan's food and agriculture system*. Working Paper No. 01-0312. Michigan State University Product Center. Retrieved March 13, 2015, from http://productcenter.msu.edu/uploads/files/MSUProductCenter2012EconomicImpactReport1.pdf

Lane, B. (1994). Sustainable rural tourism strategies: A tool for development and conservation. In W. Bramwell & B. Lane (Eds.), *Rural tourism and sustainable development: Proceedings of the second international school on rural development* (pp. 102–111). Clevedon: Channel View.

Little, J. (1999). Otherness, representation and the cultural construction of rurality. *Progress in Human Geography, 22*(3), 437–442.

McGehee, N. G., Kim, K., & Jennings, G. R. (2007). Gender and motivation for agri-tourism entrepreneurship. *Tourism Management, 28*(1), 280–289.

Michigan Apple Committee. (2016). Retrieved from http://www.michiganapples.com/

Mormont, M. (1990). 'Who is rural?' or, how to be rural: Towards a sociology of the rural. In T. Marsden, P. Lowe, & S. Whatmore (Eds.), *Rural restructuring – global processes and their responses* (pp. 21–44). London: Avebury.

Murdoch, J., & Pratt, A. C. (1993). Rural studies: Modernism, post-modernism and the 'postrural'. *Journal of Rural Studies, 9*(4), 411–427.

Ploeg, J. D. v. d., Renting, H., Brunori, G., Knickel, K., Mannion, J., Marsden, T., et al. (2000). Rural development: From practices and policies towards theory. *Sociologia Ruralis, 40*(4), 391–408.

Pollan, M. (2001). *The botany of desire: A plants-eye view of the world*. New York, NY: Random House.

Pritchard, A., & Morgan, N. J. (2001). Culture, identity, and tourism representation: Marketing Cymru or Wales? *Tourism Management, 22*, 167–179.

Short, B. (2006). Idyllic ruralities. In P. Cloke, T. Marsden, & P. H. Mooney (Eds.), *The handbook of rural studies* (pp. 133–148). London: Sage.

Shortridge, J. R. (1996). Keeping tabs on Kansas: Reflections on regionally based field study. *Journal of Cultural Geography, 16*(1), 5–16.

Weightman, B. A. (1987). Third world tour landscapes. *Annals of Tourism Research, 14*(2), 227–239.

Yates Grist & Cider Mill. (2016). About us. Retrieved November 15, 2016, from https://www.yatescidermill.com/

CHAPTER 6

Developing Social Capital in Craft Beer Tourism Markets

Susan L. Slocum

INTRODUCTION

The craft beer industry is currently undergoing extraordinary growth across the United States. In 2012, there were almost 2400 craft breweries throughout the country and the retail sale of craft beer generated $14.3 billion in sales annually (Goddard, 2013). However, the Commonwealth of Virginia, located firmly in the "Bible Belt", has traditionally regulated alcohol production and distribution, limiting the opportunities for craft beer sales (Slocum, 2015b). This changed in 2012, when the General Assembly allowed breweries to sell their product for on-site consumption, like wineries (Commonwealth of Virginia, 2012). Today, over 100 breweries are open around the state (Commonwealth of Virginia, 2015) resulting in $623 million as an economic impact for Virginia (Virginia Craft Brewers Guild, 2015).

Loudoun County, Virginia, has promoted itself as "DC's Wine Country" for many years, and the addition of craft beer to the tourism

S.L. Slocum (✉)
Tourism and Event Management, George Mason University, Manassas, VA, USA

© The Author(s) 2018
S.L. Slocum et al. (eds.), *Craft Beverages and Tourism, Volume 2*,
DOI 10.1007/978-3-319-57189-8_6

offer appears to be a natural partnership. Visit Loudoun (2014), the destination marketing organization for Loudoun County, conducted a survey to better understand the craft beer visitors and to determine if this emerging market is similar or different from the wine tourists to the area. Their results showed that craft beer tourists are generally nine years younger than wine tourists (averaging 39.5 years of age), are mostly male (66%), and are married (68%). Furthermore, while craft beer tourists reported above-average incomes, they were substantially lower than other studies of wine tourists (Alebaki & Iakovidou, 2011). Nearly 75% of the survey respondents reported visiting three breweries, 58% reported staying overnight, and spent, on average, $290 per visit. While in Loudoun County, these tourists also participated in local culinary dining (68%) and outdoor recreational activities (50%). The dilemma facing Loudoun County was whether to combine a new craft beer trail with the existing wine trail as a means to promote more tourism to the area or to start a new craft beer trail that operates independently from the wine trail. Recognizing the significant differences between craft beer tourists and wine tourists (Bateman, 2014), Visit Loudoun decided to explore a stand-alone craft beer trail.

This study was conducted to understand the levels of community support, through an assessment of social capital, existing networks, and the potential for collaboration, between craft brewers and the tourism industry. Using a qualitative, semi-structured interview approach with brewers, and a quantitative survey approach with tourism businesses, this chapter provides insight into the needs and concerns facing new entrants into the craft beer tourism market. This chapter demonstrates that craft beer is highly valued by businesses as a new tourism promotional opportunity. However, the results also show that a lack of bridging social capital between brewers and the tourism industry must be addressed before a craft beer trail can be effective.

Literature Review

Efforts to enhance tourist experiences and promote locally produced artisan products have resulted in a growth in craft beer tourism, especially in areas known for traditional beverage production (Hall & Sharples, 2003). Viewed as an alternative form of tourism, beer tourism has the potential to support the changing supply and demand needs of tourism and offers an experiential component to the tourism product (Mason & Mahony, 2011). Food and drink trails are considered a dominant form of cultural and food tourism (Ilbery & Kneafsey, 1998) and are well established throughout

Virginia. However, craft beer trails as a form of tourism development, particularly in rural areas, are under-investigated in academic literature.

A beer trail is defined as a collaboration of breweries, located in close proximity to each other, and often involves joint marketing efforts to promote beer consumption as a tourist activity (Plummer, Telfer, Hashimoto, & Summers, 2005). As a collective effort, beer trails face challenges in understanding and agreeing on the type of encounter to create for tourist (Mason & Mahony, 2011). Brewers and tourism agencies must construct an experience and target their product to a relatively unstudied demographic (Slocum, 2015b). Beer trails may feature a combination of well-established breweries, as well as talented new entrepreneurial establishments, to offer engagement and enjoyment for both beer connoisseurs and novices. Beer trails may involve self-drive routes, designated through brochures, maps, or phone apps, or may comprise more formally established packages that include transportation by bus or limousine, and often involve overnight stays in local bed and breakfast accommodations (B&Bs) or hotels. Private tours can be structured around birthday celebrations, company outings, or bachelor/bachelorette parties. These tours usually offer a tasting of select beers, brewery tours, and meal pairings.

While there has been recent growth in the study of craft breweries as an opportunity for tourism diversification, much of the literature used to support drink tourism studies comes from food tourism research. Food tourism has been approached from two different perspectives: those of the agricultural industries that promote a top-down policy agenda in support of increased incomes for rural farmers and those of the tourism industry that have evolved at a more localized level (Everett & Slocum, 2013). According to food tourism literature, both of these sectors suffer from a lack of information, collaboration, and distribution systems (Montanari & Staniscia, 2009). With the rise of craft breweries, cideries, and distilleries as a form of tourism expansion, these same issues are becoming more prevalent in recent research (Plummer et al., 2005; Slocum, 2015a), and many of the food tourism studies are providing insight into successful partnership programs.

One of the key elements of success lies in the development of social capital, especially between the diverse business structures of alcohol producers, farms, and tourism businesses. Everett and Slocum (2013) suggest that social capital is a key component of sustainability and is vital in uniting food and drink industries with tourism businesses. Social capital is defined as "networks together with shared norms, values and understandings that facilitate co-operation within or among groups" (Cote & Healy, 2001, p. 41). Figure 6.1 shows the components inherent in successful social

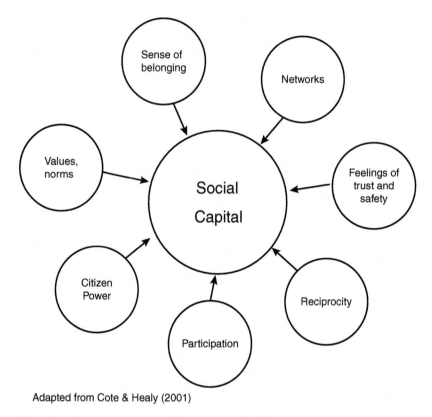

Adapted from Cote & Healy (2001)

Fig. 6.1 Elements of successful social capital building

capital building, which include networks, feelings of trust and safety, reciprocity, participation, citizen power, common values and norms, and a sense of belonging.

A high level of social capital can facilitate the flow of information, bring collective access to resources, and lower transaction costs (Alonso, 2011), all important elements of agricultural-based tourism product development. The two primary forms of social capital include bonding social capital and bridging social capital. Bonding social capital is shared networks between members of a community that share common values (Zahra & McGehee, 2013). An example would be existing partnerships between Virginia brewers or other agricultural partnerships. Bridging social capital exists between outsiders or groups with different information sources (Zahra & McGehee, 2013), such as agricultural producers and tourism businesses. Wilson, Fesenmaier, Fesenmaier, and Van Es (2001) suggest that a combination of bridging social capital and government or industry leadership provides the most effective form of social capital building and "is where sound development, management and promotion of rural tourism initiatives, such as food tourism, can best be implemented" (Everett & Slocum, 2013, p. 796).

This chapter uses information from two independent comparative case studies in an effort to provide insight into bridging social capital by highlighting the current strengths and weaknesses in social capital building between craft brewers and the tourism industry. Using Cote and Healy's (2001) model (Fig. 6.1), each element is evaluated in relation to the creation of a craft beer trail for Loudoun County, Virginia. This assessment forms the basis for a presentation of strengths and challenges facing food and drink trail development on a general platform.

Methodology

Loudoun County is segmented by two distinctly different development paths. As a suburb of Washington, DC, the eastern portion of Loudoun County possesses a highly dense, educated, population of suburban residents. These mostly professional residents make Loudoun County the second wealthiest county in the United States, with average incomes above $118,000 per year (Van Riper, 2014). Eastern Loudoun County remains rural, with 142,452 acres of farm land and an estimated 1400 farms in operation (Visit Loudoun, 2015). Loudoun County currently promotes a wine trail that includes 40 vineyards and tasting rooms and is part of the fastest growing wine-producing area in the United States (Strother & Allen, 2006). With the recent addition of 12 craft breweries, Loudoun County provides insight into the diversification process and the challenges being faced by an established drink destination that is exploring a new market for craft beer connoisseurs.

In order to understand the specific issues facing craft brewers, structured qualitative interviews were used to provide a deeper understanding of universal issues relating to the level of existing social capital, market potential, and leadership needs for a successful trail initiative (Denzin & Lincoln, 2005). Therefore, breweries were chosen from the Virginia Craft Brewers Guild website, including 11 breweries in Loudoun County and one brewery located in Arlington (another Washington, DC, suburb). Additional interviews were conducted with brewers along established beer trails around Virginia to assess potential issues in areas that receive a large percentage of tourists in order to understand concerns related to craft beer trails in more established areas. Therefore, another six interviews were conducted around the state and Table 6.1 shows the breakdown of the 17 interviews. A case study was developed using this data (see Slocum, 2015a).

An assessment of the needs, challenges, and opinions of the tourism industry was conducted using a quantitative survey that was disseminated to hotels, B&Bs, tour companies, and bus companies via email in March 2015. The hotel and B&B properties were located both in the rural and urban areas of Loudoun County and email addresses were supplied through Visit Loudoun. The tour companies were located in Washington, DC, and Loudoun County and email addresses were found through a web search of tour companies in Washington, DC, and Northern Virginia. One hundred and twenty-one surveys were sent to accommodation properties and 37 were sent to tour companies. In total, 76 surveys were completed, resulting in a response rate of 45%. All surveys were anonymous, although B&Bs were analyzed separately from hotels (based on sales volume). Table 6.2 shows general demographics of the businesses

Table 6.1 Brewery information

Brewery	Location	County
Barnhouse Brewery	Leesburg, VA	Loudon
Belly Love Brewing Company	Purcellville, VA	Loudon
Beltway Brewing Co.	Sterling, VA	Loudon
Catocin Creek Distilling Company	Purcellville, VA	Loudon
Corcoran Brewing Co.	Purcellville, VA	Loudon
Corcoran Brewing	Waterford, VA	Loudon
Crooked Run Brewery	Leesburg, VA	Loudon
Lost Rhino Brewing Co.	Ashburn, VA	Loudon
Old 690 Brewing Co.	Purcellville, VA	Loudon
Old Ox Brewery	Ashburn, VA	Loudon
Capitol City Brewing Co.	Arlington, VA	
Hardywood Park Craft Brewery	Richmond, VA	
Strangeways Brewing Co.	Richmond, VA	
Champion Brewing Co.	Charlottesville, VA	Charlottesville
C'Ville-ian Brewing Co.	Charlottesville, VA	Charlottesville
South Street Brewery	Charlottesville, VA	Charlottesville
Three Notch'd Brewing Co.	Charlottesville, VA	Charlottesville

Table 6.2 Survey demographics

Accommodation properties	Mean	Std	Min	Max
How many full-time equivalent employees do you have?	50.69	67.2	0	200
How many rooms at your establishment?	165.86	149.8	3	562
On average, how many customers do you serve in a year?	12,892	13,000	1400	6.5 million
On average, what is your annual revenue?	$250,000	214,985	45,000	650,000
We market ourselves as socially conscious or a supporter of the communities in which we operate?	58%			
Please provide the percentage of your customers that:	**Percentage**			
• Are business travelers	44.50			
• Are leisure travelers	49.09			
• Stay more than one night	53.75			
• Reside less than 100 miles from your establishment	37.11			
• Drive themselves to your establishment	70.90			
• Arrive in tour busses to your establishment	3.80			
• Are under the age of 40	42.7			
• Bring children on vacation with them	24.00			

Tour/bus companies	Mean	Std	Min	Max
How many full-time equivalent employees do you have?	22	17.9	3	79
On average, how many customers do you serve in a year?	33,157	25,402	15,000	120,000
On average, what is your annual revenue?	1,326,316	1,016,103	600,000	
We market ourselves as socially conscious or a supporter of the communities in which we operate?	12%			
Please provide the percentage of your customers that:	**Percentage**			
• Are primarily business travelers looking for an excursion on their day off	14			
• Are primarily leisure travelers	84			
• Book tours that last more than one day	12			
• Reside less than 100 miles from your establishment	24			
• Book their tour from home before arriving in your area	32			
• Are under the age of 40	34			
• Bring children on the tour with them	59			

surveyed. Seventy-three percent of the respondents represented accommodation facilities (57 responses) and 27% represented tour/bus companies (19 responses). A case study was developed using the results of this survey (see Slocum, 2015b).

This study utilizes instrumental comparative case study research in an attempt to build on existing theory and avoid theoretical assumptions about drink tourism, networking, and the development of social capital from the perspective of brewers and tourism enterprises (Slocum, Backman, & Baldwin, 2012). Two independent case studies were developed from the data sources in the spring of 2015. These independent case studies were later compared for similarities and differences which "allow(s) for the autonomy of specific cultural, social, and business networks to be reflected in the research methodology" (Slocum et al., 2012, p. 521). In particular, the goal of this research was to investigate the opinion of different actors in the potential development of a craft beer trail, including challenges, constraints, and potential resources available to brewers and tourism businesses. Together, these separate case studies allow for a more detailed investigation into social capital building.

Discussion

This section is structured according to Cote and Healy's (2001) model of successful social capital building. Each section defines the element (networks, trust and safety, reciprocity, participation, citizen power, common values, and sense of belonging) and provides evidence from the case studies related to that element. The overall discussion on bridging social capital is presented in the conclusion.

Networks

Networks allow information exchange and are a vital element when diversifying into new industries or markets. Montanari and Staniscia (2009) believe that tourism, alone, cannot increase the value of food (or drink) because they are only one element in food distribution, culinary reputation, and customer appeal. Instead, networks must occur across economic sectors to include information sources related to product development and distribution and create an integrated marketing strategy (both at the product and destination level). Everett and Slocum (2013) show that "knowledge and networking are fundamental in choosing the most appropriate vehicle (to) sustaining and growing (a) business" (p. 791), which can play a major role in the viability of food tourism. Even the very definition of social capital starts with the word "networks" (Cote & Healy, 2001).

Networking is a common theme in both case studies. Accommodations are already working closely with wineries, breweries, and local attractions (74%). While 42% are partnering with tour/bus companies, or want to expand these opportunities (80%), many accommodations feel these partnerships are one-sided (32%), where tour/bus companies rarely seek input from the properties. Those against networking claim that they are "too small to support bus trips" (43%) or that "tour companies do not currently offer overnight trips to the area" (31%). Many of the breweries are collaborating with each other as a means to promote their facility or region as a whole. Breweries along existing trails showed the highest involvement. One manager states, "we are always willing to help other breweries and feel that the rising tide of craft beer will raise all ships". This networking takes the form of sharing equipment, discussing weekly sales, and knowledge exchange for start-up brewers. However, as a new area to craft beer production, one Loudoun County brewer states, "brewers are highly competitive. It's an artisan industry, and the working relationship is somewhat non-existent in our area". As active members of Visit Loudoun, craft brewers are interested in working with tourism businesses but fear that tour companies will overwhelm their facilities. Additionally, brewers feel that networking would increase their advertising costs, which is viewed as prohibitive for start-up breweries.

Feelings of Trust and Safety

For social capital to be developed, all members of a community must feel valued and trust that their opinions matter. They must also believe that others will follow through and are genuinely concerned about the best interests of the group. Torell (2002) argues that centralized resources or uneven power distributions can erode trust. This, in turn, creates "class" favoritism (Trulsson, 1997) that can marginalize small or rural businesses in favor of urban or corporate organizations. Access to "gatekeepers" as facilitators can establish trust in competitive environments (Torell, 2002).

Power and class were easily recognizable in the independent case studies. A hierarchy exists with tour companies on top, most of which are corporately owned and headquartered in the urban centers. They tend to operate in a top-down fashion, finding attractions that will work on their terms. Corporate hotels place a close second on the hierarchy, providing brochures from local attractions but promoting activities more generally than specifically. With their captive audience, many conference centers promote wine tourism but rarely promote specific wine regions and have little preference for wineries located in their local county. Brewers and B&Bs place lowest on the ladder and are most likely to mistrust larger tourism enterprises. One brewer commented, "why would hotels and bus companies care about our business model. We have to do it their way or they will find another brewery to visit. There's no loyalty". In the quantitative study, B&Bs had the strongest belief that a craft beer trail is viable (4.75) and that there is a market for tours to rural agricultural areas in Virginia (4.75) (on a five-point Likert scale). However, a lack of resources, and the certainty that tour companies will not offer overnight stays in rural areas, has led to misperceptions about the needs and expectations of tour companies. In reality, both hotels and tour companies thought that "there is a market for craft beers" (4.3), that "there is a market for rural tours to northern Virginia" (4.0), and that "craft beer trails can be successfully marketed by tour/bus companies" (3.8). Therefore, the motivation to establish trust as a means to grow the craft beer trail exists within both the brewing and tourism industries.

Reciprocity

By its very nature, social capital is cooperation. The concept of reciprocity implies that all members are working together toward a common goal and that each member will provide their particular expertise and resources to fulfill the group's goal (Montgomery & Inkles, 2001). It is through cooperation that community members achieve certain ends that are not possible with its absence (Montgomery & Inkles, 2001). However, when one group feels that they are facing an unfair distribution of resources, or that they are required to sacrifice more than other group members, social capital breaks down. While trust is perceived (and may be earned), reciprocity usually develops over time, is embedded in traditional social networks, and is based on historical interactions between key actors (Coleman, 1988).

Because the craft beer industry is new to Loudoun County, there is a lack of participation and historical interaction between the brewers and the hospitality industry. B&Bs and breweries appeared to have the highest level of reciprocity because they are located in the same small towns and have a personal history. One brewer commented that "I went to high school with most of these guys, the owner of the winery and the daughter of the local hotel owners. I know they have my back". Another tight link was between the larger, more corporate enterprises. Sixty-seven percent of the hotels have working relationships with tour operators and offer daily tour options for their guests. It was noted that hotels and tour companies serve distinctly different markets: the hotels cater to overnight visitors and tour companies cater to day-visitors or day-trips. Therefore, there is less competition between these two industries resulting in a potentially higher level of reciprocity.

However, all groups were asked if they would participate in a joint advertising campaign to support a craft beer trail. While the brewers were generally supportive, they feared that it would be expensive. As one brewer put it, "typically the (advertising) campaigns are geared around tours that the tour company organizing makes a profit, not us". When asked if hotels would be willing to support a joint marketing campaign, 57% said yes, but none of the tour companies offered support.

Participation

Participation can be viewed as a form of equity in tourism activities and decision-making. In this instance, participation can be seen as avenues to generate revenue as a means to actively partake in craft beer tourism (often referred to as economic participation). Economic involvement can foster local control over resources, encourage participation in decision-making, and empower local people (Lele, 1991). Proximity to tourism markets can also facilitate greater participation (Goodwin, 2006). While many governing organizations are tasked with increasing participation, especially in food and drink tourism initiatives, Everett and Slocum (2013) write, "that the benefits of food tourism have not always been well communicated by governing bodies, and initiatives to encourage participation have often been met with reluctance and uncertainty as a result" (p. 799).

Both case studies showed positive support for a craft beer trail. Quantitative statements such as "craft beer trails can be successfully marketed by tour companies" (3.8) and "craft beer trails are hard to sell" (2.3) show that both hotels and tour companies are optimistic that craft beer trails add value to their product. There was also agreement that a craft beer trail should be packaged with other local attractions (3.3). Brewers felt that tourism was a positive addition to their many marketing approaches, and one brewer commented, "It doesn't hurt to bring in more people; if we were involved with tourism it would benefit us, as well as the area".

All of the case studies' participants have access to tourism markets, although those operating in urban areas have a larger potential market. Because of this, many urban businesses, especially the tour companies, feel that there is no need to participate in a craft beer trail initiative (73%). Their main explanation is that they are self-sufficient in regard to advertising, finding viable attractions, and accessing tourism markets. Accommodation properties were more inclined to participate in the development of a craft beer trail (54%) and suggest "jointly promoting winery and brewery options to encourage longer stays to the area" (25%) and "package brewery tours with overnight accommodation deals" (23%) as the best way to move forward. Breweries, on the other hand, are actively discouraging participation for fear that too many breweries along the trail would dilute the uniqueness of any single brewery. One manager estimated that "about 4-6 partners sounds about right (for a trail). Too many stops may make guests way too drunk". Currently, breweries rely on word of mouth to promote their facilities and are skeptical about too much participation and the lack of control that could result.

Citizen Power

Citizen power refers to the process of empowerment and feelings of control over resources and business decisions. Empowerment is generally tied to knowledge, as those with information can better make decisions (Coleman, 1988). The goal becomes building community capacity which provides for skill advancement and allows people to become more self-sufficient and less dependent. Businesses exist in relation to other businesses and community members, and the exercise of power and counter-power becomes a central aspect of social life (Hall, Kirkpatrick, & Mitchell, 2004). Citizen power is usually guided through effective leadership, where channels of communication allow for the flow of information and knowledge. In turn, access to this knowledge creates an even-playing field for everyone involved.

One of the issues highlighted in this research is a lack of clear leadership aspirations by any of the case studies' respondents. Although many of the breweries were intrigued by the idea of a craft beer trail, they were not interested in being the front-runners of the project. Instead, they felt that local government should provide the necessary leadership and financial resources. As one brewer pointed out, "I have a business to run. It's a great idea, but how would I fit all those responsibilities into my day?"

Tour companies and accommodation properties felt that wine and beer tourism should be promoted together (3.5) and believed that it was the responsibility of the destination marketing organization to promote the craft beer trail (3.4). As all study participants have established relationships with Visit Loudoun (and possess trust in the organization), and Visit Loudoun has access to financial resources, they appear to be the most effective leaders for the craft beer trail.

Common Values and Norms

Agriculture and tourism have very different organizational structures and value systems (Everett & Slocum, 2013), especially as it relates to food tourism promotion. Food producers (including artisan production) are often viewed as a "lifestyle business" that resist growth and increased bureaucracy (e.g., payroll and complex tax structures) (Everett & Slocum, 2013). While corporate agriculture exists, these agencies are not usually involved in the food tourism movement. Tourism can also be viewed as a lifestyle business, particularly in rural areas. However, the tourism industry also possesses a number of large, corporate entities, specifically in the hospitality, attractions, and tour operator sectors. Therefore, finding common norms and values can be more difficult in this form of niche tourism.

Community seemed to be highly valued in both case studies. Brewers repeatedly mention "business" and "community" in the same breath. For example, one participant notes, "tourism is a good development strategy, both for the business and community". This concept of community is also prevalent in the quantitative responses. When asked to rate their level of importance for "tourism that supports local communities in your area", accommodations (4.35) and tour companies (3.6) all support community values. Another common value was the visitor experience, which is rated high by hotels (4.75) and tour companies (4.6). However, "offering personalized travel experiences" and "providing small, intimate travel experiences" were more important to B&Bs (4.25, 4.45 respectively) than hotels (3.93, 3.85) and tour companies (3.6, 3.2).

A Sense of Belonging

A sense of belonging is much harder to measure than the other elements in Cote and Healy's (2001) model as it is subjective. A sense of belonging is often associated with basic human needs, like food and shelter, and can be estimated through connectedness and levels of involvement (Hall, 2014). Alonso (2011) concludes that collaboration and knowledge sharing help instill a sense of belonging in rural wine producers in the southern United States. Alonso and Bressan (2013) write, "social connectedness (is) a key element for the well-being of societies ... these are practical preconditions for safety, economic growth, and better functioning government" (p. 505). As a subjective construct, this analysis is only able to provide insight into the sense of belonging and connectedness.

At the local level, sense of belonging is evident. Brewers appear to be well integrated, both in their local community and with Visit Loudoun. This research shows active support for their cause by Loudoun County businesses and organizations. There is some collaboration with local tourism businesses, although brewers themselves are highly competitive. The interviews show a playful competitiveness, mostly protecting trade secrets and joyful boasting over regional awards. As one brewer put it, "I know (the brewer) down the road. He beat me out at the county fair last year. I like the guy, but I can't have him winning again this year". B&Bs are also willing to "actively promote activities for tourists in your local area" (100%), whereas hotels were less likely (67%). However, B&Bs were less likely to value their "working relationship between attractions" (3.75) than other hotels (4.5). Tour companies appear to be the least connected, often possessing top-down relationships with other businesses.

Conclusion

Bonding and bridging are two forms of social capital (Zahra & McGehee, 2013). While this research shows evidence of both types, the primary focus is on developing relationships between tourism businesses and craft brewers in an effort to establish a viable tourism craft beer trail. Therefore, an analysis of bringing social capital (between outsiders) is more relevant to this study. Table 6.3 shows the positive findings in relation to bridging social capital.

The highest level of social capital appears to exist between breweries and B&Bs, as many of these people have a long and established history that has led to feeling of reciprocity, networking, and a sense of belonging. As craft breweries have developed in Loudoun County, breweries and B&Bs have already established a support structure whereby breweries recommended local lodging for patrons in need and local lodging recommends breweries as an activity for their guest. However, these two groups lack

Table 6.3 Findings in relation to bridging social capital

Social capital elements	Evidence
Networks	• Accommodations and tour companies are already working closely with wineries • Accommodations already promote rural tourism • Accommodations are working with tour companies • Local B&Bs are working with local brewers
Feelings of trust and safety	• All groups trust that craft beer offers a viable diversification strategy for tourism
Reciprocity	• B&Bs and breweries have historical relationships • Hotels and tour companies have historical relationships
Participation	• All groups feel that they are well suited for economic participation
Citizen power	• Strong support from Visit Loudoun
Common values and norms	• All groups support craft beer tourism • All groups support rural tourism • All groups believe in giving back to communities
A sense of belonging	• B&Bs are the most connected with brewers • Hotels are most connected with general tourism attractions

the financial resources and time availability to partner in a more structured way (Everett & Slocum, 2013). One may argue that different members of a rural community are in fact showing bonding social capital through the promotion of their destination and the need for economic growth in their areas (Lele, 1991). Therefore, the bond (or bridge) of social connectedness between rural community players (Alonso & Bressan, 2013) is the most evident in this study.

However, the building of bridging social capital between the larger tourism system and the craft brewers is required. The uneven power distribution (caused by access to high numbers of tourist and power structures that favor larger organizations) is a very real concern on the part of brewers (Torell, 2002) and has created a hierarchy of power (Trulsson, 1997). A lack of visibility of hotels and tour companies in recommending breweries has led to a lack of reciprocal relationships (Coleman, 1988). Furthermore, tour companies are clearly mistrusted by a majority of the participants.

One of the advantages in Loudoun County is the active involvement of Visit Loudoun, which is already well connected to most of the players related to the craft beer trail. Their past research has been pivotal in understanding visitor characteristics and their leadership has already established Loudoun County as a premier wine destination. They are facilitating human capital and networking, and have the potential to build bridging social capital between these players (Coleman, 1988; Wilson et al., 2001). Also, all groups show an interest in the development of a trail, which implies that they see economic participation opportunities (Goodwin, 2006). They hold the common values of support for rural tourism and local communities. There seems to be signs that the beer trail will develop without input from the tour companies, limiting their future empowerment opportunities as they relate to craft beer tourism in Loudoun County. The outlook is optimistic for the Loudoun County craft beer trail.

REFERENCES

Alebaki, M., & Iakovidou, O. (2011). Market segmentation in wine tourism: A comparison of approaches. *Tourismos, 6*(1), 123–140.

Alonso, A. D. (2011). Standing alone you can't win anything: The importance of collaborative relationships for wineries producing muscadine wines. *Journal of Wine Research, 22*(1), 43–55.

Alonso, A. D., & Bressan, A. (2013). Small rural family wineries as contributors to social capital and socioeconomic development. *Community Development, 44*(4), 503–519.

Bateman, S. (2014). Study: Loudoun craft beer drinkers are mostly male, highly educated. Loudoun Times-Mirror. Retrieved July 30, 2016, from http://www.loudountimes.com/news/article/study_loudoun_craft_beer_drinkers_are_mostly_male_highly_educated_profes234

Coleman, J. S. (1988). Social capital in the creation of human capital. *American Journal of Sociology, 94*, 95–120.

Commonwealth of Virginia. (2012). SB 604 Alcoholic beverage control; privileges of brewery licensees. Retrieved July 2, 2015, from http://lis.virginia.gov/cgi-bin/legp604.exe?121+sum+SB0604

Commonwealth of Virginia. (2015). *Virginia craft beer month 2015*. Retrieved January 25, 2015, from https://governor.virginia.gov/newsroom/proclamations/proclamation/virginia-craft-beer-month-2015

Cote, S., & Healy, T. (2001). *The well being of nations. The role of human and social capital*. Paris: Organisation for Economic Co-operation and Development.

Denzin, N., & Lincoln, Y. (2005). *The Sage handbook of qualitative research* (3rd ed.). Thousand Oaks, CA: Sage, USA.

Everett, S., & Slocum, S. L. (2013). Food and tourism, an effective partnership? A UK based review. *Journal of Sustainable Tourism, 21*(7), 789–809.

Goddard, T. (2013). The economics of craft beer. *SmartAssets*. Retrieved January 25, 2015, from https://smartasset.com/insights/the-economics-of-craft-beer

Goodwin, H. (2006). *Measuring and reporting the impact of tourism on poverty, Cutting edge research in tourism: New directions, challenges and applications.* Guildford, UK: University of Surrey.

Hall, C. M., & Sharples, L. (2003). The consumption of experiences or the experience of consumption? An introduction to the tourism of taste. In C. M. Hall, L. Sharples, R. Mitchell, N. Macionis, & B. Cambourne (Eds.), *Food tourism around the world: Development, management and markets* (pp. 1–24). Oxford: Butterworth Heinemann.

Hall, D., Kirkpatrick, I., & Mitchell, M. (2004). *Rural tourism and sustainable business*. Clevedon: Channel View Publications.

Hall, K. (2014). Create a sense of belonging. *Psychology Today*. Retrieved January 26, 2016, from https://www.psychologytoday.com/blog/pieces-mind/201403/create-sense-belonging

Ilbery, B., & Kneafsey, M. (1998). Product and place: Promoting quality products and services in the lagging rural regions of the European Union. *European Urban and Regional Studies, 5*(4), 329–341.

Lele, S. (1991). Sustainable development: A critical view. *World Development, 19*(6), 607–621.

Loudoun County. (2014). 2014 Virginia Craft Beer Visitor Profile Report. Retrieved January 23, 2015, from https://biz.loudoun.gov/archive.aspx?amid=&type=&adid=422

Mason, R., & Mahony, B. (2011). On the trail of food and wine: The tourist search for meaningful experience. *Annals of Leisure Research, 10*(3–4), 498–517.

Montanari, A., & Staniscia, B. (2009). Culinary tourism as a tool for regional re-equilibrium. *European Planning Studies, 17*(10), 1463–1483.

Montgomery, J., & Inkles, A. (2001). *Social capital as a policy resource*. Dordecht, Netherlands: Kluwer Academy Publishers.

Plummer, R., Telfer, D., Hashimoto, A., & Summers, R. (2005). Beer tourism in Canada along the Waterloo–Wellington Ale Trail. *Tourism Management, 26*, 447–458.

Slocum, S. L. (2015a). *Easing liquor regulations in the Bible-belt: The emergence of craft beer tourism*. International Conference on Tourism. London, England, June, 2015, ISBN 978-618-81503-0-0.

Slocum, S. L. (2015b). Understanding tourism support for a craft beer trail: The case of Loudoun County, Virginia. *Journal of Tourism Planning and Development, 13*(3), 292–309.

Slocum, S. L., Backman, K., & Baldwin, E. (2012). Independent instrumental case studies: Allowing for the autonomy of cultural, social, and business networks in Tanzania. In K. Hyde, C. Ryan, & A. Woodside (Eds.), *Field guide for case study research in tourism* (pp. 521–541). UK: Emerald Publishers.

Strother, P., & Allen, R. (2006). Wine tasting activities in Virginia: Is America's first wine producing state destined to wither on the vine due to overregulation? *Thomas M. Cooley Law Review, 23*(2), 221–262.

Torell, E. (2002). From past to present: The historical context of environmental and coastal management in Tanzania. *Development Southern Africa, 19*(2), 273–288.

Trulsson, P. (1997). *Strategies for entrepreneurship: Understanding industrial entrepreneurship and structural change in northwest Tanzania, Tema Teknik och Social Förändring*. Linköpings: Universitet Linköping.

Van Riper, T. (2014, April 21). America's richest counties 2014. *Forbes*. Retrieved January 25, 2016, from http://www.forbes.com/sites/tomvanriper/2014/04/01/americas-richest-counties-2014/print

Virginia Craft Brewers Guild. (2015). Virginia craft brewers guild. Retrieved July 2, 2015, from http://virginiacraftbrewers.org/default.aspx

Visit Loudoun. (2015). Retrieved from http://www.visitloudoun.org/

Wilson, S., Fesenmaier, D., Fesenmaier, J., & Van Es, J. (2001). Factors for success in rural tourism development. *Journal of Travel Research, 40*, 132–138.

Zahra, A., & McGehee, N. G. (2013). Volunteer tourism: A host community capital perspective. *Annals of Tourism Research, 42*, 22–45.

CHAPTER 7

New Jersey Craft Distilleries: Sense of Place and Sustainability

Christina T. Cavaliere and Donna Albano

INTRODUCTION

This chapter builds upon craft distillery production in the United States by further extrapolating on the role of prohibition in the formation of entrepreneurial activities specifically in the state of New Jersey (American Distilling Institute, 2016). The nationwide ban of the production, importation, transportation and sale of alcoholic beverages known as Prohibition (Behr, 1996; Mennell, 1969) is critical to this research. This chapter examines the rebirth and evolution of distilleries in the state of New Jersey due to the recent dissolution of antiquated prohibition-era-based restrictions as a result of the enactment of new laws and licensure that allows for distilleries to produce small batch spirits in the state.

This chapter analyzes the marketing of the new evolution of seven existing distillers in New Jersey and how they are communicating sense of place leading to bio-cultural (re)development in a highly urbanized yet historically agricultural community. The laws of prohibition may have

---

C.T. Cavaliere (✉) • D. Albano
Hospitality and Tourism Management Studies,
Galloway, NJ, USA

stunted innovation in craft distillery because it formed a holding pattern in entrepreneurial and small and medium enterprise (SME) development in the state of New Jersey. Since the recent modifications to spirit production laws, along with the creation of new legislation in 2013, there has been a resurgence in micro-enterprise and craft beverage production that we argue has contributed to the (re)development of sense of place in the state. The current revival of micro-enterprise has resulted in the opening of seven distilleries that provide the opportunity for host and visitor experiences, tastings, and education. These craft distillery businesses are indeed creating web-based communication that is utilizing and expressing perceptual, sociological, ideological, political, ecological, and temporal (Cavaliere, 2017; Gruenewald, 2003) aspects of sense of place that are unique to the state. It is proposed through an in-depth content analysis of the seven existing distillery websites that the revival of craft spirit production is resulting in bio-cultural conservation. This is elucidated via the application of the Multidisciplinary Framework for Place Conscious Education (MFPCE) (Gruenewald, 2003) which outlines indicators that showcase key components of sense of place. There are epistemological inferences that underpin this study that involves the unveilings of the juxtapositions of localization versus corporatization and artisanal craft production versus homogenized manufacturing as related to the impacts on sense of place and bio-cultural conservation.

Literature Review

The history of American distilling (Huckelbridge, 2014; Rorabaugh, 1979) is a critical component of American economic, industrial, and social history. The Whiskey Rebellion (Hogeland, 2010; Slaughter, 1986), the molasses rum distillery portion of the "triangle passage" (Merritt, 1960; Ostrander, 1956), two Constitutional amendments, and the grand social experiment known as Prohibition are integral elements in the history of the United States (Kinstlick, 2011). Prohibition was the period from 1920 to 1933 when the US Constitution banned alcoholic beverages, and for a period of 13 years all liquor was made illegal. Often referred to as the "Noble Experiment" (Tyrrell, 1997), prohibition led to the first and only time an amendment to the US Constitution was repealed. Bootleggers (Rosco, 2015) violated prohibition laws and large quantities of liquor made in homemade stills supplied secret bars known as speakeasies (Gitlin, 2010; Slavicek, 2009). The Prohibition era did not simply involve the restriction of alcohol but saw changes in the history of agriculture, art, and industry. In addition, there were significant social changes including the role of women, music, and fashion (Clark, 2011; Slavicek, 2009). This chapter specifically explores the role of both the restriction and resurgence of craft spirit production in New Jersey and thus the puritanical influence on innovation in business.

Thornton (1991) highlights the social problems associated with the prohibition of alcohol sales as a crucial aspect of trade and tension with the indigenous Native American population. The prohibitionist ideology in America dates back to the Colonial period (Thornton, 1996). History states that the practical purpose of this ideology has been to secure a means of social control over the lower classes and immigrant groups (North, 1974). There may be deeper connections to the history of the puritanical legislative restrictions and entrepreneurial creativity linked to the redevelopment of sense of place in New Jersey's craft distilleries' resurgence. When the 18th Amendment (Munger & Schaller, 1997) was repealed, the sale and manufacture of liquor was once again legal.

The emergence of craft distillers has increased with the favorable changes in legislation resulting in the profusion of entrants in recent years (Kinstlick, 2011). Yet, craft distillery production in New Jersey is a very burgeoning field. In August 2013, with business and economic development in mind, New Jersey Governor Chris Christie (The State of New Jersey, 2016) enacted legislation allowing for craft distillery licenses in the state. Previously, plenary distillery licenses used by commercial spirit makers had cost $12,500 per year and had no production limits. The new law and license is $938 and allows specifically small and medium distilleries to produce up to 20,000 gallons, along with permits for distillery tours, product tastings, and sales. This was a significant change in spirit production in New Jersey as it allowed small quantities of craft production to occur which inherently allows for local production of a product that could indeed reflect a sense of place as opposed to large-scale commercial and homogenized production.

Distillers who certify that at least 51 percent of raw materials used in the distillation are grown or bought from providers in the state can label their product "New Jersey distilled" (Offredo, 2013). The new law permits liquor to be sold by wholesalers and retailers, permits distilleries to operate tours, and offers samples and sales but does not allow the sale of food or a restaurant on premises. A consumer who has toured the distillery can buy up to five liters to drink off the premises (Offredo, 2013). New

Jersey has seen the rise in the number of entrants to the craft distillery market and the impact can be seen in liquor stores, bars, restaurants, and the local communities where they reside (Kizner, 2014).

Locally distinctive foods can help to develop the image of a destination and can serve as a tourism attraction in itself (Gössling, Garrod, Aall, Hille, & Peeters, 2011). Sense of place is a multifaceted topic, a concept whose roots are derived from personal and interpersonal experiences, direct and indirect contact with an area, and cultural values and shared meanings (Farnum, Hall, & Kruger, 2005). Sense of place can be described as the entire group of cognitions and affective sentiments held regarding a particular geographic locale (Altman & Low, 1992; Jorgensen & Stedman, 2001). The study of the concept of place is at home within critical tourism studies (Blackstock, 2005; Everett, 2008; Sims, 2009; Trauer & Ryan, 2005).

The research framework that is utilized in this study supports the analysis of sense of place. It is important to note that the sensuous (Kneafsey et al., 2008) possibilities of experiences with food and foodscapes through utilization of non-representational ways of knowing are gaining critical attention in the social sciences (Cavaliere, 2017). Gruenewald (2003) details five components that contribute to a multidisciplinary framework for place-conscious education as follows: (a) perceptual, (b) sociological, (c) ideological, (d) political, (e) ecological, and Cavaliere (2017) has contributed a sixth indicator, (f) temporal, resulting from empirical research involving agritourism and climate change in New Jersey. This framework was selected and utilized in this study because it is useful in understanding subcontexts of sense of place. Gruenewald (2003, p. 622) explains that the problem is that human institutions, "such as corporations, have not demonstrated an orientation of care and consciousness toward the places that they manipulate, neglect, and destroy". The six framework indicators served to structure the website analysis of this study and also serves as the arrangement of the way in which the findings are reported below. Therefore, the selection and application of the MFPCE is well situated within this tourism research context.

Methodology

Qualitative research helps embed the field in a deeper understanding of the cultural, political, and social fiber that shapes and affects the tourism industry (Jennings, 2009; Phillimore & Goodson, 2004; Riley & Love, 2000). The global importance of tourism has generated the need for a more nuanced focus on social implications, stakeholder conflicts, economic development, environmental impact, and policy development (Phillimore & Goodson, 2004). Denzin and Lincoln (2011) support the fact that researchers deploy a wide arrange of interpretive practices in an attempt to ascertain the best understanding of their subject matter.

This chapter employs content analysis, a widely used research method for the systematic examination of communication content and performance of sample websites (Camprubí & Coromina, 2016; Krippendorff, 2004). Content analysis is a method which researchers have used to evaluate websites of tourism authorities, travel agencies, visitor attractions, and more (Law, Qi, & Buhalis, 2010). In the case of this specific study, seven websites were analyzed using a framework that contains six explicit indicators. An online search of the seven New Jersey distilleries websites resulted in a range of web content including multiple tabs of resources, links, text, images, and videos. The data were collected over a three-month period ranging from July to September of 2016. The seven websites were analyzed twice using two distinct tables created by the two authors based on the modified MFPCE framework. Each researcher analyzed each website separately and independently to partially address validity. The first analyses of the seven websites consisted of a general overview in order to identify emerging research themes as related to the MFPCE framework indicators. The researchers then met to compare the results of that first analysis which further distilled the application of the second round of specific framework indicators. This resulted in the articulation of the actual applied examples that embodied the theoretical framework indicators.

Our purpose was to focus on whether or not the distilleries communicated, via their website, a sense of place and any potential interwoven focus on the six framework indicators. Increasingly, tourists make use of websites to explore destinations they wish to visit for information and to experience a destination to a certain degree beforehand (Meintjes, Niemann-Struweg, & Petzer, 2011). Businesses, including customer-oriented and information-intensive tourism enterprises, are increasingly adopting e-business models to achieve their organizational goals (Law et al., 2010). Therefore, researchers are now able to utilize websites to download data without the need to directly engage with participants.

There are several advantages of content analysis including that it is unobtrusive, unstructured, context sensitive, and able to address large quantities of data. Content analysis examines the text, images, articles, and more of the web-based communication itself and not the individuals directly (Kim & Kuljis, 2010; Krippendorff, 2004). However, there are some limitations of this research approach which include being devoid of a theoretical basis. The authors addressed this limitation by applying a theoretical framework that allows this contribution to focuses on both what is perceivable and what is theoretically significant.

This research is positioned from a post-positivist and social constructionist perspective (Henderson, 2011). The researchers applied reflexivity by acknowledging the role of their positionality in working in their home community which allows for a more in-depth understanding of subtle socio-cultural and environmental attributes. Therefore, in supporting the current turn in post-positivist reflective research, there is a purposeful acknowledgment of subjectivity in this research. However, the application of the theoretical framework and applied indicators allows for a structuring of the researcher's interpretations (Denzin & Lincoln, 2011).

Findings

Content analysis of the seven New Jersey Distillery websites was used by employing the MFPCE framework that outlines the dimensions of place. A summary and description of the framework indicators along with examples are identified in Fig. 7.1. The MFPCE framework as modified by Cavaliere (2017) employed indicators that explored the nuanced elements of sense of place. The researchers further extrapolated upon the indicators and created applicable examples of how these can be conceptualized within distilleries. Each indicator was utilized and serves to highlight the way in which the small and medium enterprises' websites have incorporated sense of place. The following section of the chapter presents the findings in more depth using each of the six MFPCE indicators and follows with a description of how the indicators were embodied.

Fig. 7.1 James F. C. Hyde (1825–1898)

Indicator One: The Perceptual

The first indicator of the MFPCE is entitled perceptual, which identifies specific website elements that affect the five senses, including touch, taste, smell, sound, and sight. These perceptual aspects encourage visitation. Additionally, when analyzing the websites, we looked for how the spirits were packaged, labeled, and branded as part of the perceptual theme. Newsletters and other methods of communications were also considered part of the perceptual dimension. Some of the perceptual aspects of the distillery websites that correlate with the perceptual indicator are described. All but one of the New Jersey Craft Distilleries uses a New Jersey name or derivative to formally identify themselves. Cooper River, Cape May, Jersey Artisans, Pine Tavern, Jersey Sprits, and Claremont are all names that identify the distillery as having a New Jersey relationship.

Images and graphics on the distillery websites used as logos and pictures include New Jersey indigenous bridges, boats and ferries, lighthouses, and the state graphic. The bottles of distilled spirits are packaged, labeled, and branded with names and imagery true to New Jersey etymology. Examples include Cooper River Distillery's Petty's Island Rum named after Petty's Island, a 300-acre sliver of land wedged between Philadelphia and Camden in the Delaware River and is logoed with a graphic of the island. Pine Tavern's Muddy Run Jersey Style Whiskey is named after the creek that runs through the back of the distillery property. Jersey Spirits' products are named after New Jersey landmarks like Boardwalk Light Amber Rum and Barnegat White Whiskey with the label graphics displaying the beach, boardwalk, and a lighthouse. Claremont's Vodka label displays the New Jersey state graphic and their Jersey Devil Moonshine is named after and pictures the legendary creature said to inhabit the Pine Barrens of southern New Jersey.

Several of the New Jersey craft distillery websites used descriptors to communicate products like "ingredients locally sourced" from Pine Tavern Distillery's farm. They use the phrase "from our barn to your glass". Cape May's Honey Liquor includes honey harvested from local Cape May hives. Cooper River Distillery also partners with other local SMEs products to use in their spirits including grapes from a New Jersey vineyard in their brandy and beer from a Philadelphia craft brewery in their Whisky. All of the New Jersey craft distilleries communicated multiple social media resources on their websites including Facebook, Twitter, and Instagram and some included a blog, video components, photos, and the ability to sign up for a newsletter to receive additional communications. Each of the seven New Jersey craft distilleries offers and markets tours and tastings.

Indicator Two: Sociological

The second indicator of the MFPCE is entitled sociological and included elements that communicate the location of the distillery, specifically mentioning New Jersey, claims of being "first" or the "only" distillery as an identifier, the mention of folklore, gender, and nostalgia were also included. A further examination of the interplay between the New Jersey craft distilleries and how they communicate to the larger society reveals the importance of their place in this newly formed marketplace. Cape May's First Distillery is prominently communicated on their homepage. Pine Tavern boasts "Salem County's first legally operated distillery". Claremont Distillery writes "Claremont Vodka is the first and only craft spirit produced in New Jersey to be awarded a gold medal at the San Francisco World Spirits Competition, the most highly regarded spirits competition in the world". Jersey Artisan states that they are "the first distillery to open in New Jersey since before Prohibition". Lazy Eye Distillery states, "We are proud to bring you the first distilled products from Atlantic County since the time of Prohibition".

The reference of folklore speaks to the New Jersey craft distilleries' connection and acknowledgment of the culture that has shaped their industry. Jersey Artisan Distillery uses nostalgic, colonial photographs throughout their website. The Lazy Eye Distillery website included information about prohibition, New Jersey and bootlegging, the Jersey Shore, and a link to tourism including Pinelands and narratives about the Pinelands National Reserve, Atlantic County, Richland, Cape May County, and The Wildwoods.

Indicator Three: Ideological

The third indicator of the MFPCE is entitled ideological and included examining awards won and communicated for their craft spirit(s). We analyzed if the distilleries identified larger connections with the world, gender and power, humor, colonialism, and the economy. Additionally, we identified any reference to being indigenous to the product, place, or process. From an epistemology perspective, we question the deeper relationship between power dynamics in the United States case, those based upon patriarchal and puritanical linage, and how they may be juxtaposed to the role that small-quantity production of mind-altering spirits (Beveridge & Yorston, 1999; Haydock, 2015; Norlander, 1999).

Claremont Distillery identifies itself as the largest craft distillery in New Jersey. They describe their vodka as "flagship" and boast that it is the first and only spirit made in New Jersey to be awarded a gold medal at the San Francisco World Spirits Competition (the preeminent spirits competition in the world). They also communicate that it has received a 93-point rating from Wine Enthusiast Magazine. Both fermentation and distillation capacities are communicated as well as the number and size of the stills used for making their vodka. The distilleries' ability to identify or communicate any larger connection with the world from an ideological perspective is evident when they acknowledge history, ingredients, indigenous New Jersey locations and landmarks, culture, and processes.

Lazy Eye Rakii pays tribute to the "Greek Spirit" of their parents and grandparents and the values that they instilled upon them. Jersey Artisan specifies molasses from Louisiana and roasted Brazilian coffee in their Morena (Latin derivative) Rum. Pine Tavern Distillery is located on their family farm. Jersey Spirits names their spirits after New Jersey places and experiences and states, "We feel a deep respect for New Jersey and its storied landmarks and wish to preserve and honor them by integrating them within our products" (Jersey Spirits Distillery, 2016).

Indicator Four: Political

The fourth indicator of the MFPCE is entitled political and included examining all references to Prohibition and the laws that have impacted the New Jersey craft distilleries. We analyzed if the websites tested for age access or communicated the New Jersey drinking age. Additionally, we analyzed for marginality and resistance factors. We analyzed whether the shape of the state was used politically or as a geopolitical boundary identifier.

The political indicator was prominent as a majority of the distilleries not only referenced Prohibition but contextually integrated the history, laws, governmental figures, facts, and graphics into their websites. Jersey Artisan Distillery, in particular, displays a strong visual theme throughout their website with colonial graphics including colonial men fighting, photos from the 1900s of women drinking, protests, and bootlegging. Furthermore, the Jersey Artisan whiskey product (Original Sorgho

Whiskey) is named after James F. C. Hyde (1825–1898), a prominent Massachusetts businessman and political leader who had an ardent interest in botany and agriculture. According to their website, in 1857, he published the definitive work on growing, harvesting, and distilling sorghum (then called sorgho). Abolitionists and Free State farmers turned to Hyde's work and sorghum to be free of slave labor produced sugar cane. Jersey Artisan communicates beyond their distillery website with a dedicated Facebook site and website for their whisky product (http://www.jamesfchyde.com/home). The whiskey label has James F. C. Hyde's picture and signature on the bottle (Fig. 7.2).

A majority of the distillery websites comply with the Federal Trade Commission (2014) suggestion to self-regulate and require age verification before communicating with a visitor. The recommendation cites that "(when) featuring content likely to have strong appeal to minors, or that permit alcohol purchases online, consider use of age-verification technologies" (Federal Trade Commission, 2014, par. 1). Compliance to the suggestion to self-regulate speaks to the novel cultural political relationship that these new small businesses outwardly project as important. These SMEs recognized this and self-regulate in order to comply with the political landscape in which they are embedded.

Indicator Five: Ecological

The fifth indicator of the MFPCE is entitled ecological and identified all references to agricultural products including ingredients used in production, animal references, and the elements including the earth, wind, fire, land formations, weather, and seasons. Additionally, references to health, consumption, nutrition, and calories were analyzed.

The ecological theme reverberated throughout each of the distilleries' websites. Known as the Garden State, New Jersey is a leader in many forms of agricultural production (The State of New Jersey, 2016).

Fig. 7.2 Great Notch Distiller logo

Each of the seven distilleries communicated specific ingredients used in their product and distilling process. A sense of authenticity resonates as rich descriptors that are used to communicate the process such as smoky flavors, charred oak barrels and locally sourced ingredients. Although some ingredients were specifically identified as originating from outside New Jersey (molasses from Louisiana), most were referred to as "domestically grown" or New Jersey sourced including corn, honey, blueberry, grain, apple, citrus, peaches, hops, potatoes and more. Jersey Artisan Distillery specifies that their sorghum is used as a "gluten-free grain and as a fine American sugar source" (James F. C. Hyde, 2016, par. 3). Imagery used throughout several websites also reflected an ecological consciousness. For example, Pine Tavern's home page is their family-owned farm image with descriptors communicating that their line of Muddy Run Jersey Style spirits is named after the creek that runs through the back of their property. They also state that "Muddy Run Jersey Style Spirits are handcrafted in small batches using primarily local ingredients (many of which are from our farm)" (Pine Tavern Distillery, 2016, par. 4).

Claremont Distillery also displays barrel imagery and promotes an "Adopt a Barrel" program. They describe their vodka as smooth and buttery as a result of their potato fermentation and that their "small batch distillation process allows us to hand select our cuts thus capturing the cleanest parts of the distilled 'hearts' guaranteeing smoothness" (Claremont Distilled Spirits, 2016, par. 1). Lazy Eye distillery prominently communicates their spirits as "craft-certified and gluten free", thus communicating their health and nutrition consciousness as it relates to their ingredients. Jersey Artisan, in their vodka descriptors, includes the fact that it is distilled through custom copper stills and carbon filters. Their rum matures in American white oak barrels and offers hints of vanilla and caramel. Cape May Distillery's home page shows imagery of fresh herbs and fruit paired with their spirits. Verbiage includes, "Our hands-on process uses locally grown raw ingredients sourced in the state of New Jersey"(Cape May Distillery, 2016, par.1).

Jersey Spirits Distilling Company communicates a strong ecological connection boasting small batches and a handcrafted process. The heart of New Jersey is in all they do from ingredients used in the distilling process, to the reclaimed items from 1800s New Jersey farm barns used to craft their bar and tables to the names of all their products. They also commissioned a local New Jersey artist to illustrate their product labels.

Indicator Six: Temporal

The sixth indicator of the MFPCE is entitled temporal and identified specific perceptions of seasonality and time in the travel and transport of food. References to events, holidays, hours of operation and tours, age of operation, and production or historical references of the business were analyzed as related to the conceptualization of time.

Each of the seven distilleries communicated their ability to provide tours and tastings at the distilleries during a specific time. The hours varied with most open to the public on the weekends. Most welcome walkins while others request a reservation (mostly for larger parties). This was communicated in a variety of ways including dedicated tabs on their website for tours specifying their hours of operation. For example, Cooper River communicates that they host tours, tastings, and events stating that the tours are free and enticing visitors to "BYOFood" (Cooper River Distillery, 2016, par. 1). A list of upcoming events on their site (and others) indicates an aggressive off-premise commitment for the distillery aligned with their industry and brand commitment and many are tied to the time of year or season. Distillers communicate participating and hosting events including fall festivals and summer soirees. The analysis of the temporal indicator allowed for these findings to show that New Jersey SMEs could better market seasonal and holiday-based festivals and events where their products are being purveyed. There are demographic implications to the age restrictions that are legally bound to this study area. Therefore, the role of temporality, in this case age, impacts visitor demographic to this tourism product.

Seasonality and seasonal products are often examples of how the indicator of temporality surfaces in website communication for artisanal products; however, it was not prominent in craft distillery web communications in New Jersey. However, this could be an area for producers to communicate more explicitly in order to shape and further support the role and utilization of local agricultural product development. The temporal indicator was reflected in the highlighting of seasonal festivals and events. Distilleries may want to further incorporate seasonal products as part of their craft beverages. For example, Jersey Spirits Distillery makes a Jersey Summer Tea Hooch (which is a high-proof distilled spirit) as well as Jersey Apple Hooch. These products are marketed but no overt temporal marketing is evident tying the seasons to the specific New Jersey agricultural seasonal of peaches and apples products.

The temporal indicator can be further understood by examining the expression of time on these SME websites. For example, Lazy Eye

Table 7.1 Framework indicators, themes, and examples

MFPCE framework indicators	Emerging research themes	Examples
Perceptual	• The five senses	Touch, taste, sight, feel
	• Marketing	Social media, iconography, CSR
Sociological	• Nostalgia	Old timey, native New Jerseyans, Jersey Devil,
	• Identity	Gender, old world vs. new world,
Ideological	• Awards	Certifications, metals
	• Social relations	Native Americans, slavery, Africans, Europeans
Political	• Prohibition (historical)	Bootlegging, Jersey tourism, Garden State
	• Legislation (current)	State boundaries, distribution, where to buy
Ecological	• Raw materials	Corn, rye, barley, grapes,
	• Elements	Fire, reclaimed wood, water
	• Geography	Rivers, mountains,
Temporal	• "The firsts"	First distillery, first rum, biggest distributor
	• Age	Aged barrels, age of drinker,
	• Time and seasonality	Food miles, time in transport, time of tours, seasonal products
	• Local	

Distillery has two locations and they communicate seasonal hours with more availability during the summer, limited to Saturday and Sunday during the fall. They charge $10 per person for the tour which includes three tastings. Jersey Artisan Distillery requests reservations for their tours and tastings and have tour tickets available through multiple e-commerce marketplaces. They also boast public tasting schedules where they can be found participating in seasonal events and festivals. Many of the distilleries seem to engage in external/off-premise events communicated on their websites where visitors can participate in larger format festivals and tastings based on holidays and seasons (Table 7.1).

Conclusion

In summary, this study reinforces the notion that craft producers hold the power to reflect and in some cases recreate sense of place, unlike corporate conglomerates that drive globalization and cultural homogenization.

These small entrepreneurs are valuable contributors to sustainable tourism redevelopment especially in a location such as New Jersey that has historically suffered from unsustainable peri-urban corporatization that has eroded the cultural, social, and economic structures and unique sense of place. The artisanal power of the craft beverage producers analyzed in this study has been resurrected by the dissolution of antiqued restriction through the creation of new legislation. It is important to note that not every business utilized or expressed sense of place within each indicator of the MFPCE framework. Scholars could assist these businesses where they may be weak on sense of place and where they may be able to capitalize on marketing their product and uniqueness further.

This particular industry is unique in that it allows for the use of all of the elements of the MFPCE indicators to communicate sense of place. The six indicators are strongly aligned with craft beverage production that it is clear that they contribute to sense of place. Naming farms, history of politics, sourcing of local ingredients, nostalgia, local geography, and use of local folklore and iconic infrastructure (bridges) are all specific examples of how this craft industry is reinforcing sense of place in New Jersey. The entrepreneurs all demonstrated the notion that New Jersey was a significant source of inspiration and important for their marketing and demonstrated that local production does indeed communicate the importance of place. Through the intentional communication and utilization of unique elements of place, these businesses can be touted with the contribution of bio-cultural conservation. They clearly articulate how New Jersey is different from anywhere else and that uniqueness is the quintessential backbone to the essence of sustainable tourism as it counteracts the constant drive toward homogenization of place. New Jersey can be argued as the front line of capitalism, yet the examples derived from this study support how localization and artisanal and craft production reinforce both subtle and drastic senses of place and in this context and for these authors ... home.

References

Altman, I., & Low, S. M. (1992). *Place attachment, human behavior, and environment: Advances in theory and research* (Vol. 12). New York, NY: Plenum Press.
American Distilling Institute. (2016). Craft certification. Retrieved from http://distilling.com/resources/craft-certification/

Behr, E. (1996). *Prohibition: Thirteen years that changed America*. New York, NY: Arcade Publishing, Inc.

Beveridge, A., & Yorston, G. (1999). I drink, therefore I am: Alcohol and creativity. *Journal of the Royal Society of Medicine, 92*(12), 646.

Blackstock, K. (2005). A critical look at community based tourism. *Community Development Journal, 40*(1), 39–49.

Camprubí, R., & Coromina, L. (2016). Content analysis in tourism research. *Tourism Management Perspectives, 18*, 134–140.

Cape May Distillery. (2016). Retrieved from http://www.capemay-distillery.com/

Cavaliere, C. T. (2017). *Cultivating climate consciousness: Agritourism providers' perspectives of farms, food and place*. Unpublished doctoral dissertation, University of Otago, Dunedin, New Zealand.

Claremont Distilled Spirits. (2016). Retrieved from http://claremontdistillery.com/vodka/

Clark, N. H. (2011). *The dry years: Prohibition and social change in Washington*. Seattle, WA: University of Washington Press.

Cooper River Distillery. (2016). Tours, tastings & events. Retrieved from http://cooperriverdistillers.com/402-2

Denzin, N. K., & Lincoln, Y. S. (2011). *The Sage handbook of qualitative research*. Los Angeles, CA: Sage.

Everett, S. (2008). Beyond the visual gaze? The pursuit of an embodied experience through food tourism. *Tourist Studies, 8*(3), 337–358.

Farnum, J., Hall, T., & Kruger, L. E. (2005). *Sense of place in natural resource recreation and tourism: An evaluation and assessment of research findings*. General Technical Report, PNW-GTR-660. United States Department of Agriculture, Pacific Northwestern Research Station. Retrieved from http://www.webpages.uidaho.edu/css385/Readings/Farnum_Sense_of_place_pnw_gtr660.pdf

Federal Trade Commission. (2014). Self-regulation in the alcohol industry. Retrieved from https://www.ftc.gov/system/files/documents/reports/self-regulation-alcohol-industry-report-federal-trade-commission/140320alcoholreport.pdf

Gitlin, M. (2010). *The Prohibition era*. North Mankato, MN: ABDO Publishing.

Gössling, S., Garrod, B., Aall, C., Hille, J., & Peeters, P. (2011). Food management in tourism: Reducing tourism's carbon 'foodprint'. *Tourism Management, 32*(3), 534–543.

Gruenewald, D. A. (2003). Foundations of place: A multidisciplinary framework for place-conscious education. *American Educational Research Journal, 40*(3), 619–654.

Haydock, W. (2015). Understanding English alcohol policy as a neoliberal condemnation of the carnivalesque. *Drugs: Education, Prevention and Policy, 22*(2), 143–149.

Henderson, K. A. (2011). Post-positivism and the pragmatics of leisure research. *Leisure Sciences, 33*(4), 341–346.

Hogeland, W. (2010). *The Whiskey Rebellion: George Washington, Alexander Hamilton, and the frontier rebels who challenged America's newfound sovereignty.* New York, NY: Simon and Schuster.

Huckelbridge, D. (2014). *Bourbon: A history of the American spirit.* New York, NY: Harper Collins.

James F. C. Hyde Original Sorgho Whiskey. (2016). Our story. Retrieved from http://www.jamesfchyde.com/home

Jennings, G. (2009). Methodologies and methods. In T. Jamal & M. Robinson (Eds.), *The handbook of tourism studies* (pp. 672–692). Los Angeles, CA: Sage.

Jersey Spirits Distillery. (2016). Retrieved from http://www.jerseyspirits.com/about.html

Jorgensen, B. S., & Stedman, R. C. (2001). Sense of place as an attitude: Lakeshore owners attitudes toward their properties. *Journal of Environmental Psychology, 21*(3), 233–248.

Kim, I., & Kuljis, J. (2010, June). *Applying content analysis to web based content.* Information technology interfaces (ITI), 2010 32nd International Conference on (pp. 283–288). IEEE.

Kinstlick, M. (2011). *The US craft distilling market: 2011 and beyond.* Retrieved from http://crftrs.com/blog/wp-content/uploads/2014/02/craft-distillingmarket-2011.pdf

Kizner, M. (2014). N.J. craft distillery license changes provide exciting opportunity. Retrieved from http://www.nj.com/opinion/index.ssf/2014/06/opinion_nj_craft_distillery_license_changes_provide_exciting_opportunity.html

Kneafsey, M., Cox, R., Holloway, L., Dowler, E., Venn, L., & Tuomainen, H. (2008). *Reconnecting consumers, producers and food: Exploring alternatives.* Oxford: Bloomsbury Publishing.

Krippendorff, K. (2004). *Content analysis: An introduction to its methodology.* Thousand Oaks, CA: Sage.

Law, R., Qi, S., & Buhalis, D. (2010). Progress in tourism management: A review of website evaluation in tourism research. *Tourism Management, 31*(3), 297–313.

Meintjes, C., Niemann-Struweg, I., & Petzer, D. (2011). Evaluating web marketing of luxury lodges in South Africa. *African Journal of Marketing Management, 3*(9), 233–240.

Mennell, S. J. (1969). Prohibition: A sociological view. *Journal of American Studies, 3*(02), 159–175.

Merritt, J. E. (1960). The triangular trade. *Business History, 3*(1), 1–7.

Munger, M., & Schaller, T. (1997). The prohibition-repeal amendments: A natural experiment in interest group influence. *Public Choice, 90*(1–4), 139–163.

Norlander, T. (1999). Inebriation and inspiration? A review of the research on alcohol and creativity. *The Journal of Creative Behavior, 33*(1), 22–44.

North, G. (1974). *Puritan economic experiments.* Tyler, TX: Remnant Press.
Offredo, J. (2013). New Jersey distilled: Gov. Chris Christie signs craft distillery bill into law. Retrieved from http://www.nj.com/mercer/index.ssf/2013/08/new_jersey_distilled_gov_chris_christie_signs_craft_distillery_bill_into_law.html
Ostrander, G. M. (1956). The colonial molasses trade. *Agricultural History, 30*(2), 77–84.
Phillimore, J., & Goodson, L. (2004). *Qualitative research in tourism: Ontologies, epistemologies and methodologies* (Vol. 14). New York, NY: Routledge.
Pine Tavern Distillery. (2016). About us. Retrieved from http://www.pinetaverndistillery.com/our-story.html
Riley, R. W., & Love, L. L. (2000). The state of qualitative tourism research. *Annals of Tourism Research, 27*(1), 164–187.
Rorabaugh, W. J. (1979). *The alcoholic republic: An American tradition.* New York, NY: Oxford University Press.
Rosco, H. (2015). *Drinking and remaking place: A study of the impact of commercial moonshine in east Tennessee.* Unpublished Master's Thesis, University of Tennessee, Knoxville.
Sims, R. (2009). Food, place and authenticity: Local food and the sustainable tourism experience. *Journal of Sustainable Tourism, 17*(3), 321–336.
Slaughter, T. P. (1986). *The Whiskey Rebellion: Frontier epilogue to the American Revolution.* New York, NY: Oxford University Press.
Slavicek, L. C. (2009). *The Prohibition era.* New York, NY: Infobase Publishing.
The State of New Jersey. (2016). The Garden State. Retrieved from http://www.state.nj.us/nj/about/garden/
Thornton, M. (1991). *The economics of prohibition.* Salt Lake City, UT: University of Utah Press.
Thornton, M. (1996). The fall and rise of puritanical policy in America. *Journal of Libertarian Studies, 12*(1), 143–160.
Trauer, B., & Ryan, C. (2005). Destination image, romance and place experience—An application of intimacy theory in tourism. *Tourism Management, 26*(4), 481–491.
Tyrrell, I. (1997). The US prohibition experiment: Myths, history and implications. *Addiction, 92*(11), 1405–1409.

CHAPTER 8

Drink Tourism: A Profile of the Intoxicated Traveler

Kynda R. Curtis, Ryan Bosworth, and Susan L. Slocum

Introduction

Food and drink has long been regarded as an important component of the overall tourist experience, where culinary encounters provide the local cultural or authentic experience sought by tourist (Cleave, 2013; Sims, 2009). However, food and drink has now transformed from an important component of the travel experience into a primary factor in destination selection (Croce & Perri, 2010). In fact, tourists now travel to specific areas to experience the local cuisine, including drink, as evidenced by a US Travel Association report showing 17% of American travelers engaged in culinary or drink-related activities across a three-year period (Sohn & Yuan, 2013). The growing demand for food tourism, and specifically drink tourism, is

---

K. Curtis (✉) • R. Bosworth
Department of Applied Economics, Utah State University,
Logan, UT, USA

S.L. Slocum
Tourism and Event Management, George Mason University,
Manassas, VA, USA

evidenced by the increasing emergence of wine and ale trails, wine and beer festivals, and an ever-expanding number of microbreweries, wineries, cideries, and distilleries across the United States. In the Carolinas, for example, "beer pilgrimages", where travelers visit a grouping of breweries on one trip, are on the rise (Kiss, 2015).

Food and drink tourism is often used as a rural economic development strategy to turn rural communities into tourism destinations, providing unique branding and image marketing opportunities (Plummer, Telfer, Hashimoto, & Summers, 2005; Yuan, Cai, Morrison, & Linton, 2005). In primary wine-producing states, such as Washington or California, annual wine-related tourism expenditures are significant. In 2011, Napa Valley alone had wine-related tourism expenditures of $1.05 billion (Stonebridge Research, 2012). Substantial economic impacts have been noted in other states as well. In North Carolina in 2009, 1.2 million winery visits led to $156 million in wine-related tourism expenditures (Rimerman, 2009). In Michigan, the economic impacts of wine tourism were estimated at $75 million (Wargenau & Che, 2006). Beer-related tourism is also an important contributor. For example, the 2014 Oregon Brewers Festival added $32.6 million to the economy (Oregon Craft Beer, 2014) in 2011, and craft breweries generated approximately $3 billion in total economic impact in California (Richey, 2012).

Despite the importance of food and drink tourism in rural economic development and its growth in popularity among tourists, drink tourism overall has been under-researched (Hall, Sharples, Cambourne, & Macionis, 2000; Mitchell, Hall, & McIntosh, 2000). In order to understand drink tourism, it is important to understand or profile drink tourists. However, previous studies attempting to describe drink tourists have primarily focused on wine tourism, used qualitative or basic quantitative measures, and have limited their data collection to a subsection of visitors at wineries, breweries, or festivals rather than a full selection of tourists (Dodd & Bigotte, 1997; Getz, 2002; Getz & Brown, 2006; Williams & Dossa, 2003). Additionally, tourist psychographics, such as interests, values, and lifestyles, have been examined on a limited basis, with the majority focusing only on basic demographics and vacation characteristics (Charters & Ali-Knight, 2002; Mitchell et al., 2000; Sohn & Yuan, 2013).

This chapter widens the examination of drink tourism through the use of a complete data set of travelers to the US Intermountain West, rather than only wine tourists or visitors to wineries, breweries, and so on. Additionally, it examines tourists involved in many types of drink-related activities and compares them to non-drink tourists by demographic and psychographic characteristics. Finally, advanced quantitative methods are employed to profile drink tourists and factors likely to increase participation in drink tourism.

Literature Review

The literature on food tourism, which includes drink tourism, is extensive, so the discussion here focuses only on those studies surrounding drink-related tourism, and specifically those that seek to describe or profile drink tourists. The majority of literature on drink tourism surrounds wine-related regions and events. For example, Dodd and Bigotte (1997) used cluster analysis to analyze a consumer survey conducted at Texas wineries. Using demographic data on age and income, the authors constructed two consumer segments and suggested that the differences between the two groups were a result of their current stage in the family life cycle (older with grown children vs. younger with children). However, it is important to note that the focus of the study was on the purchasing of wine by tourists rather than participation in wine-related tourism.

Getz and Brown (2006) used factor analysis to examine wine tourists through a survey of wine club members in Calgary, Canada. They concluded that wine tourists were most interested in destinations with cultural and outdoor attractions, as well as a variety of activities in which to participate. A study by Charters and Ali-Knight (2002) examined tourists to two wine regions in Australia through the use of in-person winery visitor surveys. The respondents were placed into four groups (lover, connoisseur, interested, and novice) based upon respondent self-reported interest and knowledge of wine. The characteristics and interests of each group where then compared.

Williams and Dossa (2003) examined wine tourists through a survey of non-resident visitors to British Columbia, Canada (BC). The study compared wine tourists to non-wine tourists through the use of Chi-square and t-tests. Additionally, cluster analysis was used to identify wine tourist market clusters. The study found that wine tourists were well-educated, employed baby boomers seeking natural landscape and cultural/social aspects on their vacation to BC. They also visited locations they had been to before or that were recommended by friends, made travel arrangements well in advance, traveled in larger parties, and stayed in BC longer than the non-wine visitors.

Yuan et al. (2005) examined attendee characteristics and motivations at an Indiana wine festival. Study results found that although festival attendees were a very heterogeneous group in attendance for a variety of reasons, overall, they were younger, college educated, higher-income females. Attendees traveled in small groups, made the decision to attend the festival a week out, and traveled less than 30 miles to the festival. The authors attributed some of their results to the festival's urban location and noted that a rural festival location might affect attendee types.

Another study surrounding wine festivals was conducted by Sohn and Yuan (2013) at the First Annual Lubbock Wine Festival in Texas. This study examined both demographic and psychographic characteristics of festival attendees and grouped respondents into five "motivation" groups including idealist, achiever, explorer, belonger, and innovator and then made group comparisons.

Studies focusing on drink tourism, other than wine, include Plummer et al. (2005) who examined beer tourism through visitor surveys conducted at six breweries along an ale trail in Ontario, Canada, from 1998 to 2000. Survey results were analyzed through summary statistics. Visitors to the breweries were younger (74% under 40 years of age) and traveled in small groups. They traveled for one day only and several breweries were visited in that day. The "ale trail" was noted as very important in the destination selection by 46% of the respondents.

This chapter extends the literature on drink tourism through the use of survey data collected in 2013–2014 of travelers to the US Intermountain West. The analysis compares the demographic and psychographic characteristics of drink tourists to non-drink tourists through tests of differences in means. Additionally, logit and ordered logit models are used to examine the connection between tourist characteristics and their propensity to engage in drink tourism. Marketing suggestions and destination management implications are provided for major study findings.

SURVEY DATA OVERVIEW

Data were collected through in-person traveler (non-Utah resident) surveys conducted in the summer of 2013 and winter of 2014 at 12 sites in Utah, including tourist information centers in gateway cities, entrances to national parks, airports, ski areas, and convention and visitor centers. The survey was pre-tested with Utah State University (USU) alumni attending a week-long "summer college" on the USU campus in Logan, UT. Four trained surveyors wearing USU shirts conducted the surveys, each asking every third person passing by to complete the survey. Respondents completed the survey via paper or iPad, and no time limit for survey completion was given. A total of 700 usable surveys were completed across the survey period. No data was collected on those who refused to participate in the survey and no incentives, in terms of gifts, cash, coupons, and so on, were provided to encourage survey participation.

The survey included a number of questions regarding respondent socio-demographics, current trip characteristics, their food- and drink-related activities at home and while traveling, and their perceptions of and interests in various tourism activities in the Intermountain West. Sample summary statistics are provided in Table 8.1 and are grouped into several categories starting with demographics. Sample demographic results

Table 8.1 Sample summary statistics and difference in mean test results

Variable	Full sample			Non-drink tourists		High-drink tourists		Difference in means
	Mean	St. Dev.	Obs	Mean	Obs	Mean	Obs	t-test
Demographics								
Age	49.91	17.05	635	49.14	477	52.22	158	−1.97
Female = 1	0.48	0.5	668	0.46	498	0.52	170	−1.44
Married = 1	0.68	0.47	665	0.68	496	0.7	169	−0.5
High education (BA/BS degree or higher) = 1	0.68	0.47	700	0.68	510	0.68	190	−0.15
Have children = 1	10.3	0.46	700	0.32	510	0.24	190	**2.18**
Reason for visit								
Business travel = 1	0.14	0.34	700	0.12	510	0.17	190	−1.54
Visit family or friends = 1	0.05	0.21	700	0.05	510	0.04	190	0.38
Visit national parks = 1	0.08	0.27	700	0.08	510	0.08	190	0.06
Participate in outdoor rec. = 1	0.37	0.48	700	0.35	510	0.42	190	−1.66
Visit cultural/heritage sites = 1	0.21	0.41	700	0.22	510	0.18	190	0.97
Attend special event = 1	0.02	0.14	700	0.03	510	0.01	190	**1.8**
Participate in agritourism activities = 1	0.08	0.28	700	0.09	510	0.07	190	0.54

(continued)

Table 8.1 (continued)

Variable	Full sample			Non-drink tourists		High-drink tourists		Difference in means
	Mean	St. Dev.	Obs	Mean	Obs	Mean	Obs	t-test
At-home activities (5 point scale, never to always)								
Buy local foods	3.52	1.16	666	3.42	496	3.79	170	−3.56
Shop at farmers markets	3.08	1.14	667	2.93	498	3.54	169	−6.27
Participate in a CSA	1.73	1.11	627	1.62	478	2.07	149	−4.38
Buy certified organic produce	2.79	1.19	659	2.76	494	2.88	165	−1.08
Visit local farms	2.05	1.14	652	1.98	490	2.25	162	−2.67
Cook at home	4.29	0.8	665	4.24	494	4.46	171	−3.08
Try new foods/recipes	3.86	0.91	659	3.77	494	4.13	165	−4.55
Buy unfamiliar produce	2.93	1.19	659	2.8	496	3.29	163	−4.63
Eat ethnic foods	3.57	1.06	655	3.44	494	4	161	−6.06
Attend beer/wine events	2.39	1.27	658	2.1	498	3.28	160	−11.07
Food canning/preserving	1.8	1.11	652	1.77	492	1.89	160	−1.18
Beer/wine making	1.35	0.88	648	1.23	490	1.71	158	−6.09
Home gardening	2.62	1.5	656	2.52	493	2.93	163	−3.06
Compost	2.14	1.49	655	2.09	493	2.3	162	−1.53
Recycle	4.35	1.07	654	4.28	487	4.53	167	−2.6
While traveling activities (5 point scale, never to always)								
Buy local foods	2.8	1.19	620	2.7	500	3.23	120	−4.37
Shop at farmers markets	2.46	1.16	624	2.35	504	2.96	120	−5.28
Visit local farms	1.77	0.95	607	1.7	497	2.05	110	−3.56
Spend night at local farms	1.28	0.65	609	1.24	499	1.5	110	−3.92
Agritourism activities	1.6	0.88	606	1.55	499	1.84	107	−3.15
Cook at accommodation	2.82	1.31	635	2.83	507	2.78	128	0.38
Try new foods	3.48	1.06	617	3.37	500	3.97	117	−5.6
Try local recipes	3.12	1.19	610	3.01	499	3.65	111	−5.26
Buy food gifts/souvenirs	2.72	1.13	622	2.59	502	3.28	120	−6.22
Eat at locally sourcing restaurants	2.97	1.23	623	2.79	501	3.69	122	−7.52
Attend beer/wine events	2.22	1.28	620	1.75	510	4.4	110	−32.67
Recycle	3.67	1.33	622	3.59	501	4.03	121	−3.33

(continued)

Table 8.1 (continued)

Variable	Full sample			Non-drink tourists		High-drink tourists		Difference in means
	Mean	St. Dev.	Obs	Mean	Obs	Mean	Obs	t-test
Utah Knowledge/Attitudes (5 point scale, strongly disagree to agree)								
Known for outdoor activities	4.48	0.69	678	4.47	505	4.51	173	−0.68
Known for landscapes	4.63	0.61	679	4.63	506	4.64	173	−0.24
Known for heritage/culture	3.86	0.89	666	3.87	497	3.83	169	0.49
Interest in native American culture	3.65	1	664	3.63	496	3.73	168	−1.14
Interest in Mormon culture	2.73	1.27	665	2.75	499	2.67	166	0.69
Strong food culture in Utah	2.94	0.79	665	2.97	496	2.86	169	1.54
Food in Utah is good	3.88	0.74	662	3.88	494	3.89	168	−0.25
Local food advertised well	2.94	0.87	658	2.93	493	2.99	165	−0.78
Saw Utah's own/local first Utah labels/signs	2.5	1.07	655	2.53	491	2.4	164	1.34
Attractions well advertised	3.35	0.93	653	3.33	492	3.4	161	−0.76
Local crafts available	3.36	0.89	650	3.35	488	3.4	162	−0.5
Understand Utah culture	3.16	0.96	657	3.12	490	3.29	167	**−1.97**
Enough time do/see all	2.79	1.21	658	2.79	494	2.77	164	0.16
Planning to return	4.1	0.89	672	4.09	501	4.15	171	−0.81
Will recommend Utah to friends/family	4.43	0.69	672	4.37	501	4.6	171	**−3.67**

show that 68% of the respondents were married, 48% female, at an average age of 50 years. The average annual income in 2012 of the sample was $103,000, and the respondent's ethnic background included 84% white, 5% Asian, and 4% Hispanic. The respondent's educational level was high, as 28% had a bachelor's degree and 40% had a graduate degree. Half of the respondents (49%) were employed full time, while 29% were retired.

The respondent's average length of stay in Utah was 10.6 days, and they traveled in groups of 4.5 people (2.9 adults and 1.6 children) on average. One-third of respondents (30%) had children under the age of 18. The respondents were visiting Utah for a variety of reasons. The primary reason was to participate in outdoor activities (37%), followed by visiting cultural/heritage sites (21%), business travel (14%), visiting national parks (8%), and agritourism activities (8%). The internet was the most common way that respondents researched or booked their trip, followed by other (14%), and brochures (10%). Almost one-third came to Utah as an annual tradition (32%).

Survey respondents were asked to indicate how often they participate in food, drink, and related activities at home on a scale of 1–5, where 1 is never and 5 is always (whenever possible). Respondent average ratings for each activity are provided in Table 8.1. Cooking at home, recycling, and trying new foods/recipes were rated the highest. However, respondent interest in local foods, farmers' markets, and ethnic foods were also strong. Survey respondents were also asked to indicate how often they participate in a similar set of food, drink, and related activities while traveling on a scale of 1–5, where 1 is never and 5 is always (whenever possible). Ratings for attending beer/wine tastings and festivals were highest, followed by trying new foods, and trying local recipes. Cooking at their accommodations and eating at locally sourcing restaurants were also highly rated.

Lastly, respondents were asked to rate on a scale of 1–5, where 1 is strongly disagree and 5 is strongly agree, their knowledge, attitudes, and interests while in Utah. Respondents felt that Utah was mostly known for its scenic landscapes and outdoor activities, although they rated the food they consumed in Utah highly (3.88), as well as its heritage/culture (3.86). However, they did not indicate the food culture in Utah was strong and did not feel local foods in Utah were advertised well. On a positive note, their plans to return and their intention to recommend Utah to others was rated highly (4.1 and 4.43 respectively).

Analysis and Results

For the purposes of this study, drink tourism includes participation in activities surrounding beer, wine, and distilled spirits, including tastings, tours, trails, and festivals. Survey sample statistics show that almost "a fifth" (18%, 110 respondents) of the respondents participate in drink-related activities often when traveling and another 20% participate in such activities occasionally. The distribution of responses related to participation in drink-related activities while traveling is provided in Fig. 8.1.

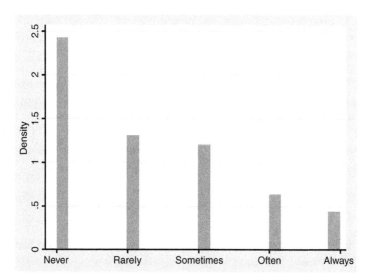

Fig. 8.1 Distribution of respondent participation in drink tourism

Comparing Drink and Non-Drink Tourists

Respondents were grouped into two sets, "non-drink tourists" or those who participate in drink tourism never, rarely, or only sometimes, and "high drink tourists" or those who participate in drink tourism often or always. The demographic and psychographic characteristics of each group are compared, and Table 8.1 shows summary statistics for each group and t-test results for differences in means across the two groups.

The results of the t-tests reveal several statistically significant differences between groups. Respondents who frequently participate in drink tourism are older by three years on average, and are less likely to have children under the age of 18. They are also more likely to report outdoor recreation as a reason for visiting Utah, but less likely to be in Utah for a special event. The group of variables that describe "at home activities" collectively indicate that high-drink tourists are more likely to be engaged in local food activities as they purchase local foods at grocery stores, farmers' markets, belong to community-supported agriculture (CSA) programs, and visit local farms. They are also more likely to seek out food experiences such as trying unfamiliar foods, new recipes, and eating ethnic foods more often. Finally, high-drink tourists are also beer and wine enthusiasts at home and thus, more likely to attend drink-related events and do their own brewing or wine making at home.

While traveling, high-drink tourists exhibit similar tendencies related to local foods such as frequenting local-sourcing restaurants and participating in farm-based hospitality and events. They are also more likely to purchase food-related gifts or souvenirs. In general, high-drink tourists do not display markedly different levels for variables that describe knowledge of and attitudes about the state of Utah than non-drink tourists. The only exceptions include their higher perceived understanding of Utah culture and their intention to recommend Utah as a destination to others.

Modeling Visitor Propensity for Drink Tourism

To better understand the connection between respondent attributes (demographics and psychographics, such as interests, values, and lifestyle) and the propensity to engage in drink tourism, respondent response to the frequency of their drink tourism participation is analyzed as the dependent variable in a set of logit and ordered logit regression models (see Wooldridge (2010) for a full description of these models). The logit models use a constructed dummy variable called "high drink", which is equal to one if the value of the initial variable is four or five (often or always) as the dependent variable. The ordered logits analyze the initial variable in its full set of ratings (1–5, or never to always). Which of these methods is preferred is unclear—the logit-based method has the advantage of identifying self-reported frequent drink tourists in a clean, discrete manner. But, the ordered logit retains the full distributional properties of the original variable and provides a look at what factors are statistically associated with an increase in drink-related tourism from, say, two to three (rarely to sometimes), or three to four (sometimes too often). These factors may provide insights into the types of activities, promotional methods, and so on, which may influence visitors to increase their engagement in drink tourism.

In Tables 8.2, 8.3, 8.4, 8.5, and 8.6, the results of both the logit and ordered logit models are presented. Each model contains a different "block" of variables following the organization presented in Table 8.1. Moreover, each model is reported both with and without the suite of demographic variables as controls. Table 8.2 shows results associated with the basic demographic variables. While none of the logit model results

Table 8.2 Logit model results—demographic variables

Variable	Logit	Ordered logit
Age	0.0099	0.0002
	(0.0064)	(0.0052)
Female	0.1786	−0.0465
	(0.1912)	(0.1554)
Married	−0.0855	0.1229
	(0.2322)	(0.1890)
High education	0.1464	0.5748
	(0.2122)	(0.1758)**
Have children	−0.2369	−0.4640
	(0.2231)	(0.1770)**
Ord. logit threshold_1		−0.0844
		(0.2856)
Ord. logit threshold_2		0.8647
		(0.2872)**
Ord. logit threshold_3		1.9728
		(0.2955)**
Ord. logit threshold_4		2.9275
		(0.3188)**
Constant	−1.7043	
	(0.3604)**	
Obs	613	558

*$p < 0.05$; **$p < 0.01$

were statistically significant, the ordered logit results were very similar to the mean difference t-test results. Those with fewer children under the age of 18 and those who had obtained at least a bachelors' degree were more likely to participate in drink-related tourism. These results for respondent demographics are consistent across the models presented in Tables 8.3, 8.4, 8.5, and 8.6. These results indicate that adults traveling with their families on vacation are likely not a target market for drink-related tourism activities or events, but rather older couples traveling alone.

Table 8.3 shows results associated with the variables describing the respondents' reasons for visiting Utah. Interestingly, there is little statistical significance among these variables. The logit model suggests that individuals in Utah for outdoor recreation or for business reasons are more likely to be high-drink tourists; however, the ordered logit model indicates a higher propensity to participate in drink tourism among those who are in

Table 8.3 Logit model results—Utah visit rational

Variable	Logit	Logit	Ordered logit	Ordered logit
Business travel	1.1789	1.2989	0.6360	0.5769
	(0.5276)*	(0.6591)*	(0.3592)	(0.3858)
Family or friends	0.7169	1.2118	0.5693	0.6910
	(0.6295)	(0.7354)	(0.4306)	(0.4503)
National parks	0.8508	1.0272	−0.6278	−0.4507
	(0.5677)	(0.7010)	(0.4308)	(0.4504)
Outdoor recreation	1.0454	1.4166	0.5496	0.6429
	(0.4993)*	(0.6264)*	(0.3208)	(0.3442)
Cultural/heritage sites	0.7021	0.9579	0.6816	0.8138
	(0.5185)	(0.6520)	(0.3331)*	(0.3634)*
Special event	−0.7828	−0.4064	−1.0342	−1.0078
	(1.1413)	(1.2079)	(0.6477)	(0.6704)
Agritourism	0.7112	1.0127	−0.2208	−0.2839
	(0.5704)	(0.6929)	(0.4033)	(0.4292)
Age		0.0109		0.0052
		(0.0065)		(0.0054)
Female		0.1836		−0.0751
		(0.1939)		(0.1586)
Married		−0.1229		0.0020
		(0.2369)		(0.1944)
High education		0.2202		0.6063
		(0.2170)		(0.1784)**
Have children		−0.1288		−0.4417
		(0.2356)		(0.1899)*
Ord. logit threshold_1			−0.0166	0.5265
			(0.2979)	(0.4244)
Ord. logit threshold_2			0.9090	1.5208
			(0.3001)**	(0.4282)**
Ord. logit threshold_3			1.9685	2.6558
			(0.3097)**	(0.4367)**
Ord. logit threshold_4			2.9926	3.6191
			(0.3311)**	(0.4540)**
Constant	−1.8563	−2.9715		
	(0.4809)**	(0.7061)**		
Obs	700	613	620	558

*$p < 0.05$; **$p < 0.01$

Utah to visit cultural or heritage sites. This may indicate that if one wishes to experience the heritage of a place, they may try local alcoholic beverages, but would not normally participate in drink tourism.

Overall, the results in Table 8.4 suggest that individuals who are involved in local food and associated food-based cultural activities at home

Table 8.4 Logit model results—activities at home

Variable	Logit	Logit	Ordered logit	Ordered logit
Buy local foods	0.1091	0.1090	0.2458	0.2149
	(0.1171)	(0.1329)	(0.0969)*	(0.1034)*
Shop farmers markets	0.3009	0.2855	0.0791	0.0200
	(0.1263)*	(0.1425)*	(0.1042)	(0.1093)
Participate CSA	0.1663	0.1114	0.2044	0.2118
	(0.1231)	(0.1371)	(0.0985)*	(0.1070)*
Buy organic produce	−0.2379	−0.1735	−0.0455	0.0021
	(0.1107)*	(0.1237)	(0.0865)	(0.0916)
Visit local farms	−0.0705	0.0203	−0.0883	−0.0350
	(0.1220)	(0.1327)	(0.0967)	(0.1038)
Cook at home	−0.0178	0.0716	−0.4071	−0.3658
	(0.1736)	(0.1934)	(0.1271)**	(0.1346)**
Try new foods/recipes	0.0638	0.0001	0.2349	0.3386
	(0.1602)	(0.1840)	(0.1271)	(0.1408)*
Buy unfamiliar produce	−0.0052	0.0650	−0.1425	−0.1344
	(0.1109)	(0.1237)	(0.0945)	(0.0998)
Eat ethnic foods	0.2798	0.3051	0.0834	0.0589
	(0.1305)*	(0.1467)*	(0.1014)	(0.1068)
Attend beer/wine events	0.6499	0.6830	1.3122	1.2879
	(0.0997)**	(0.1102)**	(0.1010)**	(0.1054)**
Food canning/preserving	−0.1096	−0.1244	0.0016	0.0192
	(0.1227)	(0.1344)	(0.0966)	(0.1025)
Beer/wine making	0.2436	0.2272	0.2731	0.2357
	(0.1257)	(0.1400)	(0.0997)**	(0.1097)*
Home gardening	0.0982	0.1030	0.1924	0.1727
	(0.0935)	(0.1017)	(0.0759)*	(0.0794)*
Composting	0.0246	0.0057	−0.0674	−0.0717
	(0.0876)	(0.0958)	(0.0695)	(0.0727)
Recycle	0.0097	0.0014	0.0342	0.0220
	(0.1163)	(0.1270)	(0.0905)	(0.0947)
Age		0.0085		0.0087
		(0.0092)		(0.0066)
Female		0.4238		−0.0532
		(0.2571)		(0.1852)
Married		−0.4015		−0.1983
		(0.3141)		(0.2280)
High education		−0.1469		0.3138
		(0.2918)		(0.2171)
Have children		−0.0314		−0.3059
		(0.2840)		(0.2079)

(*continued*)

Table 8.4 (continued)

Variable	Logit	Logit	Ordered logit	Ordered logit
Ord. logit threshold_1			3.4658	4.0619
			(0.6191)**	(0.7711)**
Ord. logit threshold_2			4.9988	5.6474
			(0.6351)**	(0.7880)**
Ord. logit threshold_3			6.5839	7.2914
			(0.6635)**	(0.8128)**
Ord. logit threshold_4			8.0630	8.6841
			(0.7060)**	(0.8499)**
Constant	−5.4334	−6.4188		
	(0.8682)**	(1.1193)**		
Obs	571	515	533	486

*$p < 0.05$; **$p < 0.01$

are more likely to engage in drink tourism. The logit model suggests drink tourists are likely to attend farmers' markets, shop for organic produce, eat ethnic foods, and attend local drink-related events at home. The ordered logit indicates that individuals who shop for local foods, subscribe to CSA programs, cook frequently at home, and actively try new foods and recipes are more likely to participate in drink-related events while traveling. The same is true for respondents who attend drink-related events at home, make beer or wine at home, and engage in home gardening.

Table 8.5 reports results for the suite of variables describing respondent activities while traveling. The logit model shows a statistically significant association with high-drink tourists and those respondents who are more likely to try new foods while traveling, eat at locally sourcing restaurants, and stay on farms. The ordered logit model identifies those respondents who do not cook at their accommodations, eat at locally sourced restaurants, and recycle while traveling as having a higher probability of participating in drink tourism while traveling. This makes intuitive sense as those who eat at restaurants frequently will have increased opportunities to sample local and regional beers, ciders, wines, and so on.

Finally, Table 8.6 shows results associated with respondent interests, knowledge, and attitudes about Utah. In general, these variables are not predictive of participation in drink tourism, with a few exceptions. The logit model suggests that those who are more likely to recommend Utah as a travel destination to friends are more likely to be high-drink tourists.

Table 8.5 Logit model results—activities while traveling

Variable	Logit	Logit	Ordered logit	Ordered logit
Buy for local foods	0.0152	−0.0667	0.0025	−0.0213
	(0.1352)	(0.1497)	(0.0882)	(0.0937)
Shop at farmers markets	0.1372	0.2328	0.0871	0.1391
	(0.1448)	(0.1607)	(0.1038)	(0.1097)
Visit local farms	−0.1221	−0.2217	−0.0401	−0.0652
	(0.1713)	(0.1871)	(0.1232)	(0.1283)
Spend night at local farm	0.4237	0.3637	0.3892	0.3066
	(0.2123)*	(0.2473)	(0.1551)*	(0.1772)
Agritourism	−0.0438	0.0289	0.2856	0.3216
	(0.1640)	(0.1804)	(0.1129)*	(0.1204)**
Cook at accommodation	−0.1995	−0.2178	−0.2122	−0.2072
	(0.1026)	(0.1180)	(0.0710)**	(0.0770)**
Try new foods	0.4976	0.7283	0.1999	0.3552
	(0.1740)**	(0.1902)**	(0.1162)	(0.1266)**
Try local recipes	−0.0212	−0.0715	0.1097	0.0646
	(0.1391)	(0.1454)	(0.1008)	(0.1055)
Buy food gifts/souvenirs	0.2349	0.1795	0.1401	0.0870
	(0.1205)	(0.1326)	(0.0832)	(0.0896)
Eat at locally sourcing restaurants	0.4960	0.5118	0.3357	0.3237
	(0.1316)**	(0.1449)**	(0.0834)**	(0.0892)**
Recycle	0.1721	0.1768	0.1914	0.1536
	(0.1047)	(0.1192)	(0.0653)**	(0.0716)*
Age		0.0040		0.0027
		(0.0096)		(0.0059)
Female		0.7677		0.2697
		(0.2869)**		(0.1754)
Married		0.0586		0.1404
		(0.3308)		(0.2078)
High education		0.2673		0.4681
		(0.3205)		(0.1981)*
Have children		−0.1677		−0.4210
		(0.3199)		(0.1938)*
Ord. logit threshold_1			3.1162	3.6879
			(0.4117)**	(0.5626)**
Ord. logit threshold_2			4.2190	4.8480
			(0.4279)**	(0.5774)**
Ord. logit threshold_3			5.4683	6.1676
			(0.4545)**	(0.6021)**
Ord. logit threshold_4			6.6969	7.3368
			(0.4891)**	(0.6333)**
Constant	−6.4326	−7.7082		
	(0.7375)**	(1.0285)**		
Obs	568	513	566	512

*$p < 0.05$; **$p < 0.01$

Table 8.6 Logit model results—Utah knowledge, attitudes, and interests

Variable	Logit	Logit	Ordered logit	Ordered logit
Known for outdoor activities	0.0654 (0.2049)	0.3881 (0.2534)	0.2137 (0.1710)	0.1922 (0.1879)
Known for landscapes	−0.1838 (0.2273)	−0.2503 (0.2669)	0.0019 (0.1816)	0.0772 (0.1950)
Known for heritage/culture	−0.0591 (0.1294)	0.0472 (0.1479)	0.0409 (0.1107)	0.0949 (0.1197)
Interest in native American culture	0.1711 (0.1116)	0.1202 (0.1222)	0.3648 (0.0936)**	0.3124 (0.0984)**
Interest in Mormon culture	−0.0747 (0.0836)	−0.0892 (0.0913)	−0.1193 (0.0713)	−0.1540 (0.0751)*
Strong food culture in Utah	−0.1500 (0.1445)	−0.2256 (0.1592)	−0.1942 (0.1315)	−0.2405 (0.1430)
Food in Utah is good	−0.0922 (0.1444)	−0.0864 (0.1602)	−0.3299 (0.1288)*	−0.3086 (0.1358)*
Local food advertised well	0.1088 (0.1447)	0.0218 (0.1587)	0.1996 (0.1302)	0.1099 (0.1368)
Saw Utah's own/local first Utah	−0.0689 (0.1080)	−0.0064 (0.1198)	0.1041 (0.0923)	0.1727 (0.0988)
Attractions well advertised	0.0378 (0.1276)	−0.0074 (0.1401)	−0.1162 (0.1107)	−0.1150 (0.1154)
Local crafts available	−0.0541 (0.1321)	−0.0469 (0.1433)	−0.0528 (0.1146)	−0.0710 (0.1202)
Understand Utah culture	0.2193 (0.1263)	0.2483 (0.1415)	0.0739 (0.1030)	0.0882 (0.1099)
Enough time to do/see all	−0.0076 (0.0842)	−0.0384 (0.0927)	0.0604 (0.0719)	0.0230 (0.0766)
Planning to return	−0.1339 (0.1349)	−0.2898 (0.1525)	−0.0554 (0.1168)	−0.0987 (0.1243)
Will recommend Utah	0.5427 (0.1914)**	0.7114 (0.2226)**	0.0728 (0.1455)	0.1131 (0.1558)
Age		0.3881 (0.0070)		0.0009 (0.0058)
Female		0.3861 (0.2192)		0.1326 (0.1693)
Married		−0.1318 (0.2663)		−0.0526 (0.2077)
High education		0.3081 (0.2562)		0.7112 (0.1994)**
Have children		−0.1115		−0.3703 (0.1891)

(*continued*)

Table 8.6 (continued)

Variable	Logit	Logit	Ordered logit	Ordered logit
Ord. logit threshold_1			0.6493	1.0023
			(0.8359)	(0.9153)
Ord. logit threshold_2			1.5746	1.9861
			(0.8381)	(0.9175)*
Ord. logit threshold_3			2.6031	3.1171
			(0.8446)**	(0.9252)**
Ord. logit threshold_4			4.1301	3.6619
			(0.9368)**	(0.8550)**
Constant	−2.6193	−4.4102		
	(1.0737)*	(1.3117)**		
Obs	597	539	554	505

*$p < 0.05$; **$p < 0.01$

The ordered logit model indicates that individuals who are interested in learning more about Native American culture or traditions and those who disagree that "The food in Utah is good" are more likely to engage in drink tourism.

DISCUSSION AND CONCLUSIONS

This study provides further insight into the types of tourists, in terms of demographics, interests, values, and lifestyles, who prefer drink-related tourism. Data were collected through in-person surveys conducted at twelve locations across Utah in 2013–2014. Data were analyzed through differences in mean comparisons between drink and non-drink tourists, as well as logit and ordered logit models seeking to explain visitor propensity for drink tourism.

Overall, study results provide evidence that drink tourism is experiential consumption, where drink tourist motivations fall in the four realms of an experience, including educational, esthetic, escapist, and entertainment (Pine & Gilmore, 1999; Quadri-Felitti & Fiore, 2012). Study results as they relate to the four realms of experience are provided below along with potential activity applications.

Drink tourists to the Intermountain West are middle-aged, highly educated, with no or few children at home. This result is consistent with studies conducted on wine tourists (Getz & Brown, 2006; Williams & Dossa, 2003), but differs in terms of age from beer tourist and food/wine festival studies (Plummer et al., 2005; Yuan et al., 2005), which find that younger (20s and 30s), educated visitors without children traveling in groups frequent ale trails and microbreweries. Results indicate that marketing drink tourism destinations as a couple's getaway, with educational activities, such as pairings or cooking classes, could be effective.

Further, study results show that those highly involved in outdoor recreation and cultural activities, with an interest in Native American culture, are more likely to be drink tourists. This is consistent with the wine tourism literature (Getz & Brown, 2006), especially with Williams and Dossa (2003) who found that wine tourists in British Columbia, Canada, often visited native cultural sites and purchased native arts and crafts. This is also consistent with Wang (1999) indicating that tourists seek authenticity. Results highlight the importance of linking drink-related activities with nearby recreational and cultural activities, creating a destination with multiple offerings. Examples might include hiking and biking trails linking wineries or breweries, ski-in distilleries, vineyard concerts, and so on.

Drink tourists were also more likely to participate in culinary activities and experience unfamiliar or new foods while traveling. The tourist's desire to experience food and drink unique to a destination has been noted in the literature (Mason & O'Mahony, 2007; Sims, 2009). This illustrates the importance of having local-sourcing restaurants nearby, holding local or ethnic food festivals, and organizing farm-to-table dinners or events.

Lifestyle and value characteristics for drink tourists included an affection for local foods (shopping at farmers' markets and CSAs), preferences for organics, and participation in home-based activities such as cooking, gardening, and beer and wine making. These results indicate the importance of a healthy lifestyle in terms of diet, supporting local farmers, and a connection to food production. These results are very similar to those found in consumer studies and the local food movement (Curtis & Cowee, 2011; Gumirakiza, Curtis, & Bosworth, 2014). These results highlight the importance of farm-based activities and accommodations, wine and beer making courses or demonstrations, and holding local food markets or events.

This chapter provides valuable understanding into the characteristics of drink tourists and offers insight into the types of activities that either draw drink tourists to a destination or support the engagement of visitors into the local food and drink movement while traveling. Using a sample of general tourists to the US Intermountain West, this research shows that drink tourism is a viable strategy for cultural and foodie regions that actively promote experiential tourism activities. Beer and wine tourists hold many of the same characteristics as food tourists, which can support regional development through festivals, farmers' markets, local sourcing in restaurants, and other innovative food tourism strategies. Recognizing these similarities can help communities grow their destination through the inclusion of drink-related establishments and educational activities.

Limitations to this study include its applicability across other regions of the United States and the use of respondent-stated preferences and behaviors versus observed. The study was conducted in Utah with visitors entering from surrounding states (CO, NV, ID, MT, etc.). The heavy prevalence of outdoor recreational opportunities in Utah, as well as its draw as a religious destination may appeal to unique types of travelers. The rather high average length of stay (10.6 days) and the prevalence of travel to Utah as an annual tradition may indicate differences in travelers to Utah versus the average American tourist.

Further research recommendations include expanding the analysis to include a national sample of travelers. This would correct for site-specific relevance issues and also allow for regional comparisons. Additionally, studies incorporating tourist origin (state or country of residence) would allow for comparisons across differing cultural and ethnic backgrounds and provide more detailed marketing implications for drink tourism operators.

References

Charters, S., & Ali-Knight, J. (2002). Who is the wine tourist? *Tourism Management, 23*(3), 311–319.

Cleave, P. (2013). The evolving relationship between food and tourism: A case study of Devon in the twentieth century. In C. Hall & S. Gössling (Eds.), *Sustainable culinary systems* (pp. 156–168). Oxon, UK: Routledge.

Croce, E., & Perri, G. (2010). *Food and wine tourism: Integrating food, travel and territory.* Cambridge, MA: CAB International.

Curtis, K., & Cowee, M. (2011). Buying local: Diverging consumer motivations and concerns. *Journal of Agribusiness, 29*(1), 1–22.

Dodd, T., & Bigotte, V. (1997). Perceptual differences among visitor groups to wineries. *Journal of Travel Research, 35*(3), 46–51.

Getz, D. (2002). *Explore wine tourism: Management, development, & destinations.* New York: Cognizant Communications Corporation.

Getz, D., & Brown, G. (2006). Critical success factors for wine tourism regions: A demand analysis. *Tourism Management, 27*(1), 146–158.

Gumirakiza, J. D., Curtis, K., & Bosworth, R. (2014). Who attends farmers' markets and why? Understanding consumers and their motivations. *International Food and Agribusiness Management Review, 17*(2), 65–82.

Hall, C. M., Sharples, L., Cambourne, B., & Macionis, N. (Eds.). (2000). *Wine tourism around the world: Development, management and markets*. London: Routledge.

Kiss, T. (2015). *Beer tourism begins to boom in Carolinas*. Retrieved from http://www.greenvilleonline.com/story/news/local/2015/03/04/beer-tourism-begins-boom-carolinas/24375547/

Mason, R., & O'Mahony, B. (2007). On the trail of food and wine: The tourist search for meaningful experience. *Annals of Leisure Research, 10*(3–4), 498–517.

Mitchell, R., Hall, C., & McIntosh, A. (2000). Wine tourism and consumer behaviour. In C. M. Hall, L. Sharples, B. Cambourne, & N. Macionis (Eds.), *Wine tourism around the world: Development, management and markets* (pp. 115–135). Oxford: Butterworth Heinemann.

Oregon Craft Beer (OCB). (2014). *The economic impact of the Oregon brewers fest is $32.6 million*. Retrieved from http://oregoncraftbeer.org/the-economic-impact-of-the-oregon-brewers-fest-is-32-6-million/

Pine, B., & Gilmore, H. (1999). *The experience economy: Work is theater and every business a stage*. Boston, MA: Harvard Business School Press.

Plummer, R., Telfer, D., Hashimoto, A., & Summers, R. (2005). Beer tourism in Canada along the Waterloo-Wellington Ale Trail. *Tourism Management, 2005*, 447–458.

Quadri-Felitti, D., & Fiore, A. (2012). Experience economy constructs as a framework for understanding wine tourism. *Journal of Vacation Marketing, 18*(1), 3–15.

Richey, D. (2012). *California craft brewing industry: An economic impact study*. Retrieved from http://www.californiacraftbeer.com/wp-content/uploads/2012/10/Economic-Impact-Study-FINAL.pdf

Rimerman, F. (2009). *Full economic impact of wine and wine grapes on the North Carolina Economy – 2009*. Retrieved from http://www.nccommerce.com/Portals/10/Documents/NorthCarolinaWineEconomicImpactStudy2009.pdf

Sims, R. (2009). Food, place and authenticity: Local food and the sustainable tourism experience. *Journal of Sustainable Tourism, 17*(3), 321–336.

Sohn, E., & Yuan, J. (2013). Who are the culinary tourists? An observation at a food and wine festival. *International Journal of Culture, Tourism and Hospitality Research, 7*(2), 118–131.

Stonebridge Research. (2012). The economic impact of Napa County's wine and grapes, 2011. Retrieved from https://napavintners.com/downloads/napa_economic_impact_2012.pdf

Wang, N. (1999). Rethinking authenticity in tourism experience. *Annals of Tourism Research, 26*(2), 349–370.

Wargenau, A., & Che, D. (2006). Wine tourism development and marketing strategies in southwest Michigan. *International Journal of Wine Marketing, 18*(1), 45–60.

Williams, P., & Dossa, K. (2003). Non-resident wine tourism markets: Implications for British Columbia's emerging wine tourism industry. *Journal of Travel & Tourism Marketing, 14*(3–4), 1–34.

Wooldridge, J. (2010). *Econometric analysis of cross section and panel data*. Boston, MA: MIT Press.

Yuan, J., Cai, L., Morrison, A., & Linton, S. (2005). An analysis of wine festival attendee's motivations: A synergy of wine, travel, and special events? *Journal of Vacation Marketing, 11*(1), 41–58.

CHAPTER 9

Craft Brewing Festivals

Zachary M. Cook

INTRODUCTION

Pennsylvania has long been a mainstay of alcohol production. As a major destination for many German immigrants, Pennsylvania became the first colony to begin brewing lager on a large scale (Bryson, 2010, p. xvi). As one of the top producers of apples in the United States, Pennsylvania has historically been, and remains, a leader in cider production. John Chapman, more popularly known as Johnny Appleseed, used Pennsylvania apple seeds to spread his enterprise further west (Pollan, 2002, p. 26). The Whiskey Rebellion proved to be one of the most challenging moments in President George Washington's presidency when western Pennsylvania farmers violently objected to the federal tax on spirits produced from corn. Pennsylvania remains at the heart and soul of any conversation on the history of alcohol in the United States (Miller, 1963, p. 157).

According to the 2014 study conducted by the Brewers Association, Pennsylvania craft breweries churn out more craft beer than any other state in the country, over four million barrels each year (Brewers Association, 2015). D.G. Yuengling & Son is the oldest operated brewery in the nation, opening in 1829 in Pottsville, Pennsylvania. While

Z.M. Cook (✉)
American Studies, Capital College, The Pennsylvania State University, Reading, PA, USA

Yuengling has vastly expanded in recent years, gaining national fame through advertising, the company manages to stay true to its principles of local being better. The iconic company only transports their beer to nearby states and had to open a brand new brewery in Florida in order to expand their footprint to the South. In juxtaposition to Yuengling, the Straub Brewery of St. Mary's in Pennsylvania has kept a relatively small production capacity and yet has flourished since 1872. The Eternal Tap at Straub's has some tourists making yearly pilgrimages to the homey operation. Both of these venerable breweries antedate the current microbrew trend, yet are positioned to benefit as many customers move away from the biggest multinational brands. Beer culture remains robust throughout the state. New beer-tasting festivals are continually being created across Pennsylvania. Some people travel hundreds of miles just to sample goods at local beer festivals.

The people who make such treks are referred to as beer tourists and many of them will fit a few craft brewery visits into their vacation itineraries. Others may plan their entire vacation around a particular brewing region. Websites and clubs have been created to better educate the public on these matters. Local beer festivals showcase nascent breweries to the public. The beer tourist travels to festivals or breweries not to become inebriated, but simply to sample the local beer and then spread their findings to other beer tourists (Beer Tourist, 2013).

Craft beer is typically much more expensive than those produced by the macro breweries and thus can be viewed as a luxury or high-end product that depends on customer curiosity and eventual brand loyalty. Craft brewers and craft drinkers form a sort of commercial symbiosis as the former seeks to satisfy the desire of the latter for new drinking experiences. Pennsylvania's repressive beer laws, holdovers from Prohibition and aimed at protecting large beer distributorships rather than encouraging new businesses, are second only to those of Utah in terms of complexity (Levy, 2016). The laws allow only distributors to sell, and mostly by the case, although in recent years a select few grocery stores have found loopholes through which to sell six-packs. As of this writing, Pennsylvania's laws are extremely confusing. There is not only some pressure for change but also a great deal of resistance by rent-seeking distributorships and their legislative allies. Changes are slow and fitful (Levy, 2016). Beer tasting festivals allow patrons to first taste and then hopefully buy the beer at a later date.

Scholarship

Festivals have been a relatively neglected subject for folklorists and ethnographers, although that is changing. Even more neglected, because it is an entirely new form of festival, is scholarship on beer festivals. To the author's knowledge, this is uncovered territory. In this limited scholarship, research focused on festivals and their implications on society. Through his work, Pieper (1973) dealt with the idea that festivity is essentially a celebration of normal life. Pieper purported that festivals can only occur when there is a regular working day, especially within the context of a religious basis. In 1983, Hobsbawm and Ranger tackled the origins of festivals, asserting that festivals and the traditions that begin them are based on "socialization, the inculcation of beliefs, value systems and convention of behavior" (p. 9). Giorgi, Sassatelli, and Delanty take a similar track in 2011 by analyzing the place of arts festivals in the overall public domain, and Moeran and Pedersen (2011) incorporate the economic as well as cultural angle of festivals in society. Bronner (1996) studied wooden chain carvers and their connection to tradition. "Carving reinforced [the carvers'] identities and renewed their self-worth" (p. 69). In a similar manner, drinking traditionally brewed beer allows festivalgoers to feel as if they are drinking in the same manner as their ancestors, which reinforces a sense of community and belonging. Yet, this work is the first to analyze the socioeconomic status and cultural beliefs of the attendees of beer festivals.

Throughout this chapter, the terms macro and craft are used extensively and must be defined. "The macro or mass-producing sector of the [brewing] industry consists of large-scale brewers of traditional American lager beer" (Tremblay, Iwasaki, & Carol, 2005, p. 307). Macro is the beer most commonly found served in America after Prohibition, such as Budweiser, Miller-Coors, etc. (Bamforth, 2009). Microbrewing is an older term in which brewers use smaller batches in a variety of styles. "All brewers with microbrewery origins have come to be called specialty or craft brewers" (Tremblay et al., 2005, p. 308). While the craft beer market is increasing, it is still far smaller than the macro beer market (Brewers Association, 2015).

Methodology

Due to the dearth of beer festival scholarship, the methodology is based on a combination of Yin (2014) and Bronner (1996). As this is a new area of research, utilizing the basic research methodology of Yin coupled with the ethnographic and folklorist works of Bronner, a new model was created. The researcher went to craft beer festivals, saw broad access to subjects and sustained ethnographic interviewing, and thus developed a short questionnaire. Insofar as there is no standard model extant for the study of beer festivals, the following was developed: while at the beer festivals, casual conversations were held in public spaces without asking for names. Every conversation was prefaced with the following statement: "Hello, I am _____ and a student at Pennsylvania State University. Could I please ask you three questions about craft beer for research in a scholarly book?" Typically, conversations were held to less than five minutes and asked three general questions in a manner similar to that outlined by Yin, specifically the usage of survey interviews in a case study (Yin, 2014, pp. 111–112). First, why are you here? This question is designed to extract how they heard about the event and for what exact reason they decided to attend. Second, how far did you travel to get here? The distance or time traveled to attend the event will signify whether the patron needs a hotel room, plans on supporting local restaurants, or otherwise be a typical tourist in an unfamiliar area. Third, how does craft brew fit into your lifestyle? If the participant was willing, this question prompted the patron to discuss what role craft beer plays in their life, their socioeconomic status through what type of work that they do, and how successful the craft beer vendors will be at attempting to solicit additional sales through the event. Answers were recorded in paper journals by the researchers. Through these questions, statistical trends were documented as well as personal stories of the patrons of craft beer festivals in a manner similar to the ethnographic research of Bronner (1996).

Beer Festivals Overview

Three different beer festivals were visited, located in the greater southeastern region of Pennsylvania. These three sites were chosen in order to achieve a wide demographic cross section of information. One, in Philadelphia, is a relatively new festival. A second, in Oaks, has established itself as one of the longer running and more popular in the region. The third, in York, has recently attempted to reinvent itself by catering to a more upscale crowd. Each of the locales sits in a distinct cultural region of the Commonwealth.

The District 9 dance club hosted the Philadelphia Winter Beer Fest in November 2015. Besides the typical breweries lined up in aisles along what is usually the dance floor, there was a small food court that offered typical bar food such as sandwiches, French fries, wraps, etc. Two separate cigar vendors competed for business and several smaller companies selling bottle openers, hot sauce, pretzels, and more general merchandise were almost as numerous as the craft brewing companies.

Valley Forge is near Philadelphia, but very accessible to other portions of the state by virtue of its location at a major turnpike intersection. It is also the site of major shopping complexes, as well as the famous Revolutionary War historical site. In short, Valley Forge is a touristic hot spot. The indoor facility was at least five times the size of the aforementioned nightclub in Philadelphia. With concrete floors, 30-foot ceilings, and uniform white tables, the entire establishment had a polished appearance. White plastic tablecloths covered every table and appropriate bunting for the event, charities, breweries, or vendors festooned the facility.

With more than 70 craft breweries on hand and close to 20 cideries, the Valley Forge Festival clearly had far greater variety than the Philadelphia event. There were nearly as many non-beer vendors as craft breweries on hand. In this sense, the festival was reminiscent of others, such as local harvest fairs. There were merchants selling beef jerky, cigars, television streaming, liquor-infused cakes, specialty mustards, bacon, nuts, green energy, home food delivery, hot sauce, "brewscuits" for dogs, general event merchandise, and more. These sellers were attracted by the prospect of crowds and contributed to the overall commercial buzz.

In addition, the charitable aspect of the event shone through with the Committee to Benefit the Children's presence. According to Mike Marchese, board member of CBC, this suburban Philadelphia beer fest is a major fundraiser in which direct donations are acquired, raffles are drawn, and more through which research and expenses for patients with blood diseases are helped. The altruistic nature of this clearly helps the overall image of the event. This creates trust between the producer and consumer, which leads to increased consumerism (Kolm & Ythier, 2006). Pennsylvania law requires all beer festivals to have some charitable aspect, but not all events bring it to the forefront the way Starfish Junction Productions, the company that runs the Valley Forge festival, has done. According to the owner, the event's repeat customer success is derived from some or all of these factors; all contribute to a professionally run beer festival endorsed by Andy's company (M. Marchese, personal interview, December 15, 2015).

The Hibrewnation beer festival took place at York Expo Center on February 20, 2016. York, in south-central Pennsylvania, has a long history of manufacturing, which it utilizes for touristic purposes through the popular "factory tour" scene. It is in some ways a typical Rust Belt small city. York is the home of several snack food companies, a Harley-Davidson plant, and sits on the highway between Harrisburg, 25 minutes away, and Baltimore, 40 minutes away. It is also near Lancaster, 30 minutes away, a city with which it maintains a spirited rivalry; York being the White Rose City, Lancaster the Red Rose City, in a tip of the cap to the War of the Roses. Rich agricultural lands, including Amish regions, surround York. In walking into the York Expo Center, the extremely large floor concrete floor space with whitewashed walls and high ceilings seemed similar to the Valley Forge venue.

Despite his busy schedule, Matthew Davis, the director of Hibrewnation, allowed a short interview for about 15 minutes in the hour before his festival began. Davis originated Hibrewnation four years prior. Davis ran different wine and beer festivals for the past nine years and was still attempting to solidify his product so as to best please the customer and the vendors at the event. Previously, Hibrewnation was more popular, but had lost some customers to other beer festivals that started recently in the greater Harrisburg area. Davis decided to redefine the event this year by attempting to bring in a more upscale crowd and "get rid of some of the riff-raff" that had caused some minor issues in previous years. In order to do this, Davis raised his prices from 25 dollars for a three-hour session to 40 dollars, since this is the amount that the other two beer festivals had charged. In addition, Davis added a "VIP" event that included the general admission to the beer festival as well as two hours of additional tasting time in which some breweries had limited specialty craft brews available for tasting that were not accessible to the general admission crowd. The VIP patrons, amounting only to 300 people, were able to peruse the event for two hours before the more significant crowd arrived while also enjoying other perks such as beer floats and timed-release offerings. Davis's goal this year was to draw 1300 patrons. There was a charitable aspect to Hibrewnation as well. A percentage of their proceeds went toward the Sons of the American Legion for this year; in the past, the event had benefited other types of medical research. After the interview, the event was perused prior to the VIP opening in order to better observe the vendors and craft breweries available (M. Davis, personal interview, February 20, 2016).

There was roughly the same number of craft breweries attending this festival as at Valley Forge. However, the amount of non-brewery vendors present at the festival was simply astounding in both scope and volume. Besides the food court area available in the foyer, there was a fudge stand selling 30 different types of products, a bakery vendor for those with a sweet tooth, a general-purpose snack stand that gave samples of spicy ghost pepper pretzels, no less than two specialty beef jerky stands selling a wide array of flavors, as well as many others. Further vendors included bathroom remodeling, window replacements, organic soap, and a female-oriented protection company hawking stun guns and pepper spray. The local and organic stands and a spice company make holistic sense as the craft beer industry is about enjoying local, quality products. Most of these companies were available to view and had links to their own sites on Hibrewnation's website, and as such, the patrons knew that they would be present. It is doubtful that many patrons, if any, bought a ticket to the event in order to meet with these vendors. Instead, this speaks to the idea of creating a more upscale event with patrons that have more disposable income available than at the other beer festivals that I researched.

General Data Analysis from All Festivals

Sixty patrons out of approximately 300 in attendance participated in the study at the Philadelphia beer festival. Most of the attendees appeared to be between the ages of 25–50, but there were certainly many outside of this spectrum. The patrons also entirely came as either couples or groups, hardly anyone came alone to this event. Almost everyone was willing to share their status as craft beer festival tourists. In fact, they seemed pleased to be told that they represented a demographic niche subject of academic inquiry.

The Philadelphia Winter Beer Fest had the fewest number of travelers come from long distances to arrive here. Only 16% of questioned patrons had traveled more than 25 miles to participate in the event. Of the three festivals visited, this was the lowest with Valley Forge and Hibrewnation at 23% and 22%, respectively. This likely can be attributed to a number of factors, including the newness of the festival, relatively small size of the venue, and its urban as opposed to suburban location. Still, there is no great statistical difference between the three events. Approximately 20% of patrons coming from long distance are considerable. However, the vast majority of those that did so were staying with friends or family locally.

As such, the usage of hotels would be practically nonexistent for this beer festival and indeed all of the beer festivals visited. While the restaurant industry might see a relatively minor boon to this festival, the overall obvious effect on the local economy is negligible (Table 9.1).

By far, the highest percentage of festivalgoers indicated that they went to the Valley Forge Festival due to their prior participation and positive experiences during the previous years. More than 90% of the patrons assessed indicated that either they were personally at this beer fest in recent years or another member of their party had been and informed them of its geniality and quality, insisting that they come along for the next event. Thus, word of mouth is an effective method of drawing attendees to Valley Forge. Only 23% of the study population claimed to have traveled 25 miles or further to attend (Table 9.2).

With upward of 2000 attendees, the Valley Forge beer festival also had the highest resident participation of the three events; 60% of the patrons considered themselves local (see Table 9.1). Studying the vast suburban population available with easy access to roads and a gigantic parking lot,

Table 9.1 Distance traveled to reach beer festival (rounded to nearest percent), $N = 236$

104.419 pt	Philadelphia Winter Beer Fest	Valley Forge Beer and Cider Festival	Hibrewnation at York
>50 miles	3%	1%	7%
25–50 miles	13%	22%	15%
Within 25 miles	38%	15%	37%
Local	45%	60%	41%

Table 9.2 How Patrons heard about the beer fest (rounded to nearest percent), $N = 197$

	Philadelphia Winter Beer Fest	Valley Forge Beer and Cider Festival	Hibrewnation at York
Radio	5%	0%	0%
Magazine	10%	0%	0%
Billboard	3%	0%	3%
Word of mouth	33%	0%	10%
Been at festival before	10%	92%	53%
Internet/social media	40%	8%	34%

the Greater Philadelphia Expo Center at Oaks is the best geographic venue of the three events. The Valley Forge event also had more of a dance music atmosphere, thanks to a disk jockey named Meatball, sponsored by Sly Fox, a successful brewery located in nearby Pottstown. The presence of popular music further added to the festive air. Throughout the four-hour beer-tasting session, patrons had the opportunity to go onto a 30 × 20 feet dance floor and learn new dances through the use of wireless headphones. This gave the impression that when there were lulls in the ambient music the people appeared to dance without music at all, which was quite comical to passers-by. This kind of event, also called a "silent disco", is popular in such progressive venues as Portland, Oregon, and Austin, Texas (Ngun, 2012), as much because of the spectating as the participation. The quality of the venue coupled with the professionally run event makes the Valley Forge Beer & Cider Festival an event worth coming back to year after year as it has successfully enticed repeat patrons in its eighth annual incarnation. In a manner similar to Moeran and Pedersen's (2011) *Negotiating Values in the Creative Industries*, craft beer *festivals* provide more than "simply places in which to conduct business". Rather, festivals often contain a carnival-like atmosphere (Moeran and Pedersen, 2011, p. 5).

Analyzing the data from the discussions with patrons at Hibrewnation allowed the ascertaining of approximate socioeconomic status of the attendees. Doing this gave harder evidence as to whether Davis's desire to shift to a more upscale beer festival was successful or not. Categorizing people's professions into socioeconomic categories can present some challenges. Since the vast majority of participants would be unwilling or uncomfortable with sharing how much money they make, research was done under the assumption that people of similar professions make about the same amount of money and thus are in the same category as their counterparts under the full realization that a new lawyer may make a paltry fraction of one who is a partner in an established law firm, for example (Bok, 1993). For simplicity's sake, three primary categories, two minor categories, and three additional concurrent categories with the first five were created for the beer festival patrons. The three primary categories include the "prestigious", "middle class", and "working class". For the prestigious category, included were doctors, lawyers, engineers, architects, business executives, and other typically high-paying professions. The middle-class category became in some ways a catchall as both white- and blue-collar jobs were lumped together. Teachers, nurses, dental hygienists, and advertising salespeople coming led with carpenters, plumbers, and electri-

Table 9.3 Socioeconomic status of patron (rounded to nearest percent), $N = 189$

	Philadelphia Winter Beer Fest	Valley Forge Beer and Cider Festival	Hibrewnation at York
Unemployed	0%	3%	1%
Retired	2%	3%	2%
Homemaker	0%	1%	2%
Student	11%	0%	4%
Working class	0%	10%	8%
Military	13%	0%	0%
Middle class	41%	58%	42%
Prestigious	33%	25%	41%

cians. The working-class category involved the patrons whose jobs did not require much if any additional training or education and were generally considered to be low paying such as sales clerks, order selectors in a warehouse, and bartenders. Additionally, a category for military personnel and another for students was included. In addition, three other categories were included that exist as part of the other five categories. Retired beer festival patrons were included in the category of their career profession from which they retired. The few unemployed patrons were positioned in their most recent job (Table 9.3).

ANALYSIS OF SOCIOECONOMIC STATUS FROM ALL FESTIVALS

The Hibrewnation beer festival had the highest percentage of those with prestigious jobs with 41% of the overall attendees. This is a major statistical distinction wherein the Valley Forge event had 33% prestigious and Philadelphia had only 25%. However, the numbers tend to even out when one includes the middle-class category. Hibrewnation had 42% of its patrons considered to be middle class, while Valley Forge and Philadelphia had 58% and 46%, respectively. When socioeconomic status is considered simply as those who have significant disposable income (prestigious and middle class combined), then Hibrewnation's attendees comprised 83% in that category compared to Valley Forge's 83% and Philadelphia's 74%. However, the research intended to go beyond the careers of the patrons and discover if those attending the events are truly knowledgeable beer tourists participating in a social event, if they are simply a bit curious, or wanted to drink alcoholic beverages with their friends.

ANALYSIS OF CRAFT BEER LIFESTYLE FROM ALL FESTIVALS

The patrons' answers to how craft beer fit into their lifestyle allowed for a wide variety of responses and significant statistical data under the "Opinions on Craft Beer" chart given in Table 9.4, a breakdown of how the patrons felt about craft beer can be easily determined. These sets of responses were divided into four categories. The responses given are generally consistent across the research of all three festivals. The first category is "only craft" wherein the patron will eschew any type of macro offered to them and therefore will only drink beer from craft breweries or home brew. The second and third categories are "more craft than macro" and "more macro than craft", both of which are rather self-explanatory. The final category is "craft only on special occasions". If patrons fall into this category, they rarely buy craft beer and the beer festival at which they are participating might be the first time that they have tried craft. This extremely small percentage of patrons, about 2%, only came due to the celebration of a special event such as a birthday or anniversary.

Considering the fact that all of these questions were asked of patrons at a beer festival, the results of the questions were unsurprising. Approximately half of all respondents drank more craft beer than macro. Additionally, 29% said that they refused to drink macro beer. Only 19% answered that they typically drank more macro than craft beer and only 2% rarely ever drank craft beer. Perhaps more telling than the statistics is the vociferous language utilized by those who fall into the "always craft" crowd, the second most popular category. At Valley Forge, two men who fell into this prestigious category declared, "Variety is the spice of life" and as such

Table 9.4 Opinions on craft beer (rounded to nearest percent), $N = 196$

	Philadelphia Winter Beer Fest	*Valley Forge Beer and Cider Festival*	*Hibrewnation at York*
Craft only on special occasions	3%	0%	4%
More micro than craft	22%	18%	18%
More craft than micro	52%	59%	43%
Always craft	23%	23%	35%

could never demean themselves by drinking homogenous macro beer. They both came looking to find a beer that they could truly enjoy so that they could purchase it at their local beer distributor. Three other men falling into the middle-class category at the same event said that they drink craft beer because they "care about taste" and "have a soft spot for local beer, especially mom and pop" craft breweries. Another man falling into the middle-class category stated that it was much easier to avoid macro beers now that craft has become much more accessible and widespread. The harder part, he contends, is avoiding the "faux micros" in which a large, traditional beer company also produces some other flavors that they deem as craft. Inasmuch as craft beer is traditional, this is reminiscent of a general trend toward handcrafted items across a variety of artisanal fields (Seaton, 2014).

At the beginning of Hibrewnation, one attendee loudly denounced all macros by saying "I don't drink shit beers". Instead, he was out to support the little guy and will only buy American beer. Another group of men and women stated that "Everyone is drinking craft beer", indicating that craft beer's popularity continues to rise. A significant number of these same people implied or stated that they would not be drinking to excess at this event since it would defeat their primary purpose, which was to taste as many beers as possible. Most of these people either had a pad and pen or, more frequently, used a phone app such as Untappd in order to rate beers that they tasted so that they could later purchase the beers that they most enjoyed. In this way, the hobby shows itself to be systematized and an organized part of their lives, rather than merely a happenstance outing. These men and women who attempt to sample and document their opinions are beer tourists in the purest sense and truly provide an outlet for craft brewers to allow others to experience their product and potentially gain customers in the long run. They are reminiscent of other activity-based subgroups who pursue their pastimes in an organized way, and who alter their consumer habits based upon their enjoyment and experiences (Miller, 2007).

Kristyn Dolan, an event manager at Starfish Junction Productions, discussed what makes the Valley Forge festival unique and successful. Kristyn works primarily with cideries and attempting to promote them through various festivals and events. Kristyn noted that there are more cideries this year than in the past since "some people are really excited about cideries", especially females. The allure of cideries for many people is multifaceted. Besides being an extremely common colonial drink, twenty-first-century consumers relish cider's gluten-free aspect. Pennsylvania's vast apple orchards make the state a prime location for craft cideries and a greater variety of sweet and bolder styles are available to tempt a diversity of palates. The increasing popularity of cideries plays a large role in the success of the Valley Forge Beer and Cider Festival (K. Dolan, personal interview, December 15, 2015).

Brewers' Viewpoints on the Festivals

Notably, gaining additional customers is a primary reason that craft brewers participate in beer festivals and allow customers to taste their products. They are putting themselves and their products before a promising public in hopes of attracting consumers and bettering their bottom lines. They are often moved by their passion to go into the business, and the chances to reach customers motivated by similar beer passion are of obvious benefit. The goals of the craft brewers were discussed with their employees at the festivals. One of the assistant brewers for Boneshire Brewing Company of Harrisburg at Hibrewnation in York summarized what small craft breweries were looking for at this event. Boneshire had not yet opened their doors to the public (plans are set to do so in August 2016), but came to allow customers to sample their small batches of test craft beer in order to begin to get potential clientele excited about their future opening. Boneshire's goal is to produce "blue-collar" beer that appeals to everyone except those that simply love standardized, macro-produced light lager. The men and women that perform manual, skilled labor is still the segment of society into which craft beer has not fully penetrated (Ball, 2011). The participation of both the upstart craft brewer and the knowledgeable beer tourist are important to both groups as the former wants brand recognition in an ever-crowded marketplace in order to ensure future success and the latter is endeavoring to discern the best way to spend their craft beer and cider dollars.

Gender

One of the more significant, yet too rarely addressed, questions of craft beer festivals and the role that gender plays in attendance and participation. Of those patrons that participated in discussions with the researchers, it should come as no surprise that there were more men than women (Herz, 2016). Despite the stereotypes that persist of beer drinking as a masculine activity, the numbers were closer than gendered assumptions might indicate. In short, women like beer, are experienced at judicious purchasing in a competitive sales environment, and are thus obviously likely to be a major part of the craft beer scene. Though the percentages were closest at the Philadelphia beer festival, the trend of a large proportion of females falling just short of rough parity remains similar. Out of approximately 250 patrons identified by statistical analysis, 62% were male and 38% were female. Within this approximately 100 adult female segment, a rather interesting cross section of society was represented. About a quarter of the women who gave their reasons for attending the event were there because a male in their life wanted to attend and needed company. However, there were a significant number of younger women who were there for an enjoyable experience. Four professional women at the Philadelphia Winter Beer Festival noted that this was a relatively inexpensive way that "gets you drunk", but of course it also broadens your horizons to new types of beer. One of them was going to be married in the next six months and had decided to use craft beer as a theme at her wedding. Three undergraduate students at local colleges stated that they essentially came to this beer festival since they are "white and bored" and have extra money that they are willing to spend. According to these students, beer festivals have become a destination for those with nothing better to do. One of the women interviewed at Valley Forge specifically mentioned that she likes to try different ciders and that both she and her husband specifically make beer a part of their itinerary on vacations as any nearby craft brewery will likely tempt them to visit.

One female home brewer came to receive new inspirations with her own homemade creations. Her story reminds us that women frequently carry the role of shopping and preparing food at home and that home brewing could be a logical extension of the expertise thereby gained. There certainly are a large segment of females who attended the event that were far more knowledgeable than their male counterparts in the traditionally masculine field of beer. It should also be noted that at the Valley Forge Beer and Cider Festival, as well as Hibrewnation at York, there was at least one stand of wine for tasting as well and the selling of bottles. Female patrons overwhelmingly dominated the crowds at these stands. While it is statistically accurate that many of the women at the beer festival were there because their boyfriends or husbands needed someone to attend the event with them, it is important not to dismiss all females in this category as a growing number of women are becoming increasingly important in the craft beer industry as well-informed consumers.

Conclusion

Throughout this fieldwork and compilation of statistics, clear answers were found to most of the questions initially posed. In general, beer festivals are not significant to local economies in the traditional idea of using hotels and restaurants. However, the patronization of the many vendors at the festivals needs further exploration. Questions could have been formulated to elicit answers from vendors, too. A disproportionately large percentage of those beer festivalgoers have prestigious careers and an even larger number are members of the middle class. Considering the much higher cost of craft beer compared with macros, this makes logical sense. At the moment, craft beer sampling, purchasing, or brewing is not a poor person's pursuit. It has more in common with "foodie culture", in which gourmet foodstuffs attract habitués interested in the higher-quality or rare experiences of fine cuisine (Garrett, 2015). There is indeed a strong association between a higher socioeconomic status and their opinion on craft beer. Higher ticket prices and VIP options create a type of niche tourism through the affluent. Coupled with potentially strong local vendor sales, this has great retail implications for the local economy. Some of the beer festivals' attendees travel a significant distance to attend beer events, but over half of the study's participants are repeat customers to their local events.

In general, beer festival directors desire to get as many well-behaving, responsible beer tourists to their festivals as possible, give a significant portion of the proceeds to a charitable cause, and of course attempt to maximize their profits along the way. Those with greater incomes invariably are more likely to be craft beer drinkers (Watson, 2013). The vast majority of those attending beer festivals are "more craft than macro" and "always craft" drinkers and they attend mostly with a desire to sample the offerings in an attempt to find the next case of beer or cider that will ultimately end up in their home. The verbosity of the craft beer supporters indicates a strong and growing undercurrent of the counter-revolution against macro breweries in America. While Anheuser-Busch and InBev seem to be fiscally sound, an ever-increasing percentage of their profits is being taken by the craft beer industry (Brewers Association, 2015) and this trend is heavily supported by beer tastings at craft beer festivals. Finally, while mostly men attend these festivals, women also attend in remarkably large numbers and many are found to be extremely knowledgeable craft beer tourists. The vocal portion of these craft beer supporters jubilantly denounced America's still popular uniform style of light lager.

REFERENCES

Ball, D. (2011). The craft beer craze. *Sarasota Magazine, 34*, 74. Retrieved from http://ezaccess.libraries.psu.edu/login?url=http://search.proquest.com.ezaccess.libraries.psu.edu/docview/902687169?accountid=13158

Bamforth, C. W. (2009). *Beer: A quality perspective.* Burlington, MA: Academic.

Beer Tourist. (2013). *Beer tourist.* Retrieved from http://beertourist.us/

Bok, D. C. (1993). *The cost of talent: How executives and professionals are paid and how it affects America.* New York; Toronto: Free Press.

Brewers Association. (2015). *Craft beer sales by state.* Retrieved from https://www.brewersassociation.org/statistics/by-state/?state=PA

Bronner, S. J. (1996). *The carver's art: Crafting meaning from wood.* Lexington, KY: University Press of Kentucky.

Bryson, L. (2010). *Pennsylvania breweries* (4th ed.). Mechanicsburg, PA: Stackpole Books.

Garrett, R. L. (2015, August). Fill the foodie culture. *Airport Business, 29*, 20–24. Retrieved from http://ezaccess.libraries.psu.edu/login?url=http://search.proquest.com.ezaccess.libraries.psu.edu/docview/1708156082?accountid=13158

Giorgi, L., Sassatelli, M., & Delanty, G. (2011). *Festivals and the cultural public sphere.* Abingdon, Oxon: Routledge.

Herz, J. (2016, August 15). Today's craft beer lovers: Millennials, women and Hispanics. Brewers Association. Retrieved from https://www.brewersassociation.org/communicating-craft/understanding-todays-craft-beer-lovers-millennials-women-hispanics/

Hobsbawm, E. J., & Ranger, T. O. (1983). *The invention of tradition.* Cambridge: Cambridge University Press.

Kolm, S., & Mercier Ythier, J. (2006). *Handbook of the economics of giving, altruism and reciprocity* (1st ed.). Amsterdam, London: Elsevier.

Levy, M. (2016, May 29). Beer: The fight over where you buy it in Pennsylvania. *The Washington Times.* Retrieved from www.washingtontimes.com

Miller, J. C. (1963). *The Federalist era, 1789–1801.* New York: Harper & Row.

Miller, T. (2007). *Cultural citizenship: Cosmopolitanism, consumerism, and television in a neoliberal age.* Philadelphia: Temple University Press.

Moeran, B., & Pedersen, J. S. (2011). *Negotiating values in the creative industries: Fairs, festivals and competitive events.* Cambridge: Cambridge University Press.

Ngun, R. (2012, February 17). Shhhhh! Austin's first silent disco sneaks up, gets down. *Culturemap Austin.* Retrieved from http://austin.culturemap.com/news/innovation/02-17-12-the-first-noiseless-party-in-downtown-austin/

Pieper, J. (1973). *In tune with the world: A theory of festivity.* New York: Harcourt, Brace & World.

Pollan, M. (2002). *The botany of desire: A plant's-eye view of the world.* New York: Random House.

Seaton, C. T. (2014). *Hippie homesteaders: Arts, crafts, music and living on the land in West Virginia*. Morgantown: West Virginia University Press.

Tremblay, V. J., Iwasaki, N., & Carol, H. T. (2005). The dynamics of industry concentration for U.S. micro and macro brewers. *Review of Industrial Organization, 26*(3), 307–324. doi:10.1007/s11151-004-8114-9.

Watson, B. (2013). The demographics of craft beer lovers. Brewers Association. Retrieved from https://www.brewersassociation.org/insights/demographics-of-craft-beer-lovers/

Yin, R. K. (2014). *Case study research: Design and methods*. Los Angeles: Sage.

CHAPTER 10

(Micro)Movements and Microbrew: On Craft Beer, Tourism Trails, and Material Transformations in Three Urban Industrial Sites

Colleen C. Myles and Jessica McCallum Breen

CRAFT BEER IN PLACE

In 2014, the craft brewing industry contributed $55.7 billion to the United States economy (Brewer Association, 2014). Craft beer and microbreweries, defined as producing less than 6 million barrels of beer

---

We the authors would like to thank the research participants in all three sites for their cooperation as well as their delicious beer. We would also like to acknowledge research assistance provided by Garrett Wolf in Manchester, UK. Finally, we would like to extend our appreciation to the organizers and editors of the volume—Carol Kline, Sue Slocum, and Christina Cavaliere—for their support through the process of preparation and review of this manuscript.

C.C. Myles (✉)
Department of Geography, Texas State University,
San Marcos, TX, USA

J.M. Breen
Department of Geography, University of Kentucky,
Lexington, KY, USA

(Brewer Association, 2014; Herz, 2016), have become essential elements of many local and regional economies (Patterson & Hoalst-Pullen, 2014a). However, the impact of the burgeoning craft brew scene found in cities across the United States and around the world is more than just economic; breweries and the beer they produce serve a multiplicity of social purposes as well. Just as a food product may be defined by its place, so too a place may become defined by the food products it produces (Trubek, 2008). Brewing and beer have been strongly tied to place (Mittag, 2014; Patterson & Hoalst-Pullen, 2014b) due to several factors: beer flavor is heavily reliant on local water sources and the vagaries of endemic yeast strains and needs to be consumed fresh, requiring a geographically convenient population of consumers. While modern packaging advancements and commercial shipping have allowed beers to travel further, the tie to place is ever present.

A simple stroll through the aisles of a craft beer shop makes the links between beer and place legible. The propensity for craft beer shops to organize their stock by place of origin (as is often seen with wine), rather than by the type of beer, hints at the idea of terroir and the role that place and the consumption of place has in the craft brewing industry (Yool & Comrie, 2014). We also find the theme of place echoed in the names of breweries, the beers they produce, and in their packaging, where maps and landscape imagery abound (Schnell & Reese, 2003). Beer and brewing is thus a vehicle for neolocalism and the pursuit of "sustainability", in its various forms (Hoalst-Pullen et al., 2014; Holtkamp, Shelton, Daly, Hiner, & Hagelman, 2016; Schnell & Reese, 2014).

Beer tourism is established or emergent in a variety of places, including Brazil (Bizinelli, Manosso, Gândara, & Valduga, 2013), South Africa (Rogerson, 2015), and Canada (Plummer, Telfer, Hashimoto, & Summers, 2005). The use of "trails" in beer tourism, a form of cultural tourism popular in the wine tourism industry, has been on the rise (Neister, 2008). The trails take on a variety of formations with some being led exclusively by the brewers themselves to others having leadership from tourism boards or civic government (Neister, 2008). Alonso (2011), taking an "entrepreneurial" perspective of beer tourism development in Alabama, argues that locally brewed beer and food can serve as a developmental catalyst when regulation is not too restrictive. Similarly, Fastigi and Cavanaugh (2017) describe growing artisanal brewing sector in Italy, a craft industry being driven mainly by a focus on (neo)localism, noting "the current Italian artisanal beer boom as a case of reinventing tradition and as an emergent form of neolocalism, in which local forms of production and culture are articulated through global hierarchies of taste and value". As such, the effect of these trails is to formulate a cultural experience around visiting breweries and consuming a local product—beer—that can be used to focus tourism planning efforts.

With this curious combination of factors and opportunities in mind, we sought to systematically investigate how craft breweries are serving as change agents in processes of development in a variety of contexts. Utilizing an ethnographic-style, participatory approach that emphasizes the importance of fieldwork (de Wit, 2013), that is, meeting, tasting, and touring with brewers in situ, we approached multiple sites where breweries were serving these seemingly transformative roles for local communities. We also conducted secondary data analysis and review of promotional materials, popular media reporting, and other public documents investigating the relationship between brewers/breweries and processes of (re)development and/or community change.

Three Cases of Breweries as Located Within Beer Tourism Trails

Bike Dog Brewing (West Sacramento, CA)

The Sacramento region of California has over 50 breweries (either established or establishing), with much of that growth occurring in the past 5–10 years (Robertson, 2016). West Sacramento, just across the river from the state capitol, Sacramento, has played a role in building the region's identity as a beer destination while also working to establish its own beer identity and economic foundation. The West Sacramento beer scene is centered around a light industrial area of the city next to the Port of Sacramento. Three separate entrepreneurial endeavors—Bike Dog Brewing Company, Jackrabbit Brewing Company, and Yolo Brewing Company—create an ad hoc coalition of breweries, which, while not yet formally named, boasts three craft breweries within walking distance of each other. Although colloquially dubbed "Port Brewery Row", the owners of breweries note that the name is not official.

Bike Dog Brewing, the oldest of the three, is a "nano" brewery founded by three men (who happen to also be local land-use and environmental professionals) who got their start as home brewers. After strong performances at small-scale marketing efforts, they sought to expand their production. After launching a successful crowdfunding campaign, the brewery became a reality. The success of Bike Dog, in some sense, paved the way for other breweries to establish themselves in the city more broadly and in that zone of the city in particular. The two other breweries nearby, taken together with Bike Dog, have amounted to a small, but significant, community change. By repurposing a light industrial zone close to urban center(s) for both production and consumption, these small businesses have capitalized on a complimentary and compatible use in order to enhance the economy of the city as well as for their own businesses. This is an outcome mainly attributable to "coopetition" (Larsen & Hutton, 2011), wherein stakeholders in a particular industry or community experience and enact simultaneous cooperation and competition, to create synergistic community-building effects. Indeed, the local craft beer scene is involved in several kinds of community-wide initiatives, including being a factor (and actor) in establishing Sacramento as "America's Farm-to-Fork Capital" (Farm-to-Fork, 2016). Moreover, the existence—and success—of Bike Dog and the other breweries on "Port Brewery Row" has instigated other breweries to explore this zone of West Sacramento and other areas of the city as a site for other similar businesses (Robertson, 2016).

Currently, the breweries on this informal tourism trail in West Sacramento exist in the liminal regulatory space of being "industrial" food producers in an industrial area, which also happen to have the legal right to sell (up to a certain capacity) at the "farm gate" (as it is called in other (rural) direct marketing schemes). Being located in an industrial area has its perks: low rents, access to transit/shipping routes, and, in this instance, proximity to a large metropolitan area. Moreover, as the direct sales hours (evenings and weekends) are the opposite of the operating hours for the other nearby industrial uses, there is often ample parking available and little conflict with the neighbors. However, being located in an industrial area also has its detriments, namely a lack of pedestrian infrastructure, poor street lighting, and an ongoing presence of large commercial vehicles coming in close contact with a sometimes-inebriated, private-vehicle operating public. So, while informal at this time, the brewers on the nascent trail agree that formalizing their status in some way would likely be beneficial.

While there may be palpable benefits, there is the potential for conflict should the ad hoc coalition become more formally legitimized. For example, should the city officially recognize the district or commercial zone as explicitly drawing customers and tourists, that recognition might trigger more regulatory attention, both (perhaps unwanted) attention to business owners as well as attention to city liability and/or responsibility for pedestrian infrastructure needs, land use and zoning issues, and overall compatibility checks. So, at this time, remaining an ad hoc, informal, social grouping—and thus enjoying exclusively community-driven notoriety as a "trail"—is both the status quo and the preference for Bike Dog and other brewers in the area.

West Sixth Brewing (Lexington, KY)

Aimed at capitalizing on the popularity of the nearby Bourbon Trail, Kentucky's Brewgrass Trail consists of a collection of eight breweries and associated "hop spots", where craft beer can be purchased, marketed by VisitLEX, the Lexington Convention, and Visitors Bureau. Of the eight breweries, six of them are located in Lexington, while the other two, Beer Engine and Rooster Brew, are located in nearby Danville and Paris, respectively. The six Lexington breweries represent a cross section of the brewing industry. Alltech Lexington Brewing and Distilling, the oldest of the group and the only one to bottle their beer, is a subsidiary of Alltech, a livestock and poultry feed supplement producer. In addition to their Kentucky-themed beer line, they also produce a line of bourbons and provide the singular overlap between the Brewgrass Trail and the Bourbon Trail. Also included on the Brewgrass Trail are Blue Stallion Brewing, who specializes in authentic German lagers and British ales; Chase Brewing, a nano brewery housed in a former garage that serves as home to one of Lexington's two pedal pub party bikes; Country Boy Brewing, Lexington's second oldest microbrewery who recently expanded their distribution into Ohio, Indiana, Tennessee, and West Virginia; Ethereal Brewing, who focuses on Belgian farmhouse beers and conducted a crowd-sourced project collecting wild yeast from around Kentucky; and West Sixth Brewing.

West Sixth Brewing is the second largest brewery in Lexington. They gained notoriety in the craft beer community after a trademark dispute with Magic Hat that they ultimately lost. Located in a former Rainbo Bread Factory situated on the northern edge of residential Lexington, and adjacent to the historical African American hamlet of Smithtown, the brewers of West Sixth found themselves with a surfeit of space when they moved into the building and have chosen to rent the additional space to a collection of nonprofits and civic-minded businesses. The current collection includes an aquaponic farm, a coffee roaster, a bourbon distiller, the community bike shop, artist studios, a co-working space for nonprofits, and practice space for the local roller derby team. The aquaponic farm recently completed a successful crowdfunding campaign to add a teaching kitchen to the space, focusing on food access and education as well as job training. The brewery is active in supporting local charities and civic organizations with meeting space as well as with financial backing. They also host a weekly yoga class and a popular running club, which draws participants from across the city.

However, the brewery's presence in the neighborhood has not been without conflict. The large influx of people from outside the neighborhood visiting the brewery, particularly during the yoga class and running club, has led to conflicts with neighbors over already limited parking. Residents of the largely African American neighborhood have complained about rising property values pricing them out of their homes and the brewery's overwhelmingly white clientele making them feel unwelcome in their own neighborhood (Spriggs, 2014). In addition, citing suspicions of drug dealing, the brewery purchased and closed the last corner market in the neighborhood, leaving residents of the neighborhood, those with the lowest incidence of car ownership in Lexington, with nowhere to buy diapers or toilet paper within a half mile. Three years later, the building still stands abandoned.

While you can find Hop Spot stickers in the windows of Lexington bars and restaurants and you can get your Brewgrass Trail Passport stamped at any of the breweries along the trail, the Brewgrass Trail Facebook account has been deleted and its Twitter account is no longer active. The trail's webpage on the VisitLex website is still operable, and, importantly, the idea of the Brewgrass Trail still lives, even as the entity no longer appears to be actively marketed. That said, the slow fade of the Brewgrass Trail is not an indication that the project has been unsuccessful. Indeed, successful beer trails have been shuttered in other places (Plummer, Telfer, & Hashimoto, 2006). Rather, it speaks to the difficulties of cooperation among competing breweries. Hall (1999) notes that without equity among partners, collaboration is likely to fail. The distribution of resources, both financial and cultural, among the breweries on the Brewgrass Trail is decidedly uneven. Alltech Lexington Brewing and Distilling, as a subsidiary of a multi-million dollar agro-industrial corporation, has access to financial resources far beyond those of any other local brewery. As the smaller breweries on the Brewgrass Trail move forward, time will tell whether (and how) they leverage their individual resources, perhaps reorganizing into a more equitable collaboration, or if they will attempt to follow the example of West Sixth and seek the creation of (respective, solo) cultural capital.

Beer Nouveau (Manchester, UK)

Beer Nouveau in Manchester, UK, is one of an astounding 84 breweries in Manchester and is one of six breweries located on the "Piccadilly Mile" in the city. The Piccadilly Mile is a loose collection of breweries occupying a light industrial zone "under the arches" in Manchester, which is seeking to build mutually beneficial tourism and consumer opportunities among producers in this geographic zone. One of the flagship breweries on the Piccadilly Mile is Beer Nouveau, a brewery owned and operated by a former home brewer turned entrepreneur (a common trope in craft beer). The owner/brewer at Beer Nouveau claims that his brewery—and the numerous others cropping up in Manchester and across the country—is, at least partially, a response to the corporatized, homogenized, "heartless, soulless pubs" in Britain. In other words, although pub life is an essential part of British culture, the corporatization of those community institutions has made the experience less authentic, appealing, and tasty. In the brewer's words: "As long as the new, small breweries help each other out, we are pushing the large, older breweries who produce crap beer out" (Brewer, Manchester, 20 February 2016). Beer Nouveau specializes in "lost" recipes, trying to recreate, as accurately as possible, historical and sometimes forgotten beer styles, signaling that fermented beverage homogenization and the cult of "new" is not enough to satisfy discerning customers—or brewers themselves.

The Piccadilly Mile, much like Port Brewery Row and the Brewgrass Trail, is a forum for coopetition (Larsen & Hutton, 2011), a place for collaboration and an opportunity to foment a "rising tide raises all boats" situation, whereby all stakeholders gain from mutual support and cooperation. As seen in the previous cases, this can work out better or worse for those involved over time. However, the case of Beer Nouveau and the Piccadilly Mile specifically highlights how beer trails/tourism capitalize on the desire of "beer geeks" to have something (i.e., specialty beer) that is not available elsewhere and, which, moreover, counters the homogenized, corporate beer culture that is prevalent in the United States and United Kingdom. In addition, Beer Nouveau provides a prime example of the concept of "productive leisure" (De Solier, 2013) and, when expertise and a market is acquired, how home brewers become entrepreneurs (Holtkamp et al., 2016), leading to shifts in both the economy and the community.

Discussion

Craft breweries produce an identity-laden product that boosts local and regional economies, while often "taking industrial, kind of secondary, forgotten properties…and transform[ing] the neighborhood[s] around them" (Brewer, West Sacramento; 23 June 2015). However, the introduction of brewing into these post-industrial sites is not without negative consequences for the surrounding communities. In a pattern repeated in city after city, "craft beer production and consumption are used to aestheticize the industrial past and pacify resistance to…gentrification" (Mathews & Picton, 2014, p. 1). In the three cases presented here, brewery facilities inhabit previously economically marginal properties and leverage comparatively low rents to make profitable the production of a value-added product. This process, in these places, simultaneously stimulates local business through the production and distribution of the product, as well as creating a multiplier effect that sparks opportunities for economic growth and development. Specific forms of increased economic opportunity include the development of local and regional urban tourism and direct-to-consumer marketing and sales.

Beer trails, a kind of fermented tourism that can be more or less sanctioned, that is, official or unofficial, but which builds on (and/or creates) cultural cachet and identity based around locality, "craft", and unique place features. Even when these trails are in some sense imagined—that is, are fully or partially unrealized—the (nascent) idea creates energy and synergy among entrepreneurs and consumers. The breweries and trails presented here serve as catalysts and drivers of (re)development in (industrial) fringe zones of urban spaces, repurposing those places for consumption—of beer, of place—by different kinds of people. Specifically, the development of entrepreneurial and tourism opportunities encourages different people (i.e., consumers, tourists) to use spaces that were previously only used by producers (and/or residents, in case of Lexington).

In the process, place evolution stemming from the growth of breweries and associated tourism, gentrification often follows. This can be seen as either economic development or a detrimental social process. In addition, as the uses and user groups of these areas expand, the infrastructure and regulatory needs and expectations of these places and businesses also expand. Municipalities are then faced with the dilemma of authorizing or otherwise legitimizing these simultaneously productive and problematic spaces/places. But breweries, as pursued drivers of economic development, and trails, as either explicitly or implicitly sanctioned local and regional tourism engines, continue to emerge because trails build synergy, promoting coopetition (cooperative competition) between related businesses while they become tourism destinations, which bring wider economic multiplier effects, and allow for the development/gentrification of depressed or otherwise challenged urban areas.

Significantly, these economic and social changes occur without the usual pushback from existing residents and interested stakeholders (Mathews & Picton, 2014). In this way, locality and place are used to craft an experience for insiders and outsiders, burgeoning as an economic sector as residents of post-industrial countries, like the United States, continue to value experiences over stuff.

Conclusion

The three examples presented herein demonstrate that not only can breweries/brewers be change agents in and of themselves, when banded together collectively, even if informally, their influence becomes even stronger. These (micro)movements of collective action allow for cooperation even in the face of competition. This kind of coopetition (framed by Larsen & Hutton, 2011) is especially applicable in the case of burgeoning brew scenes. Similarly to emerging wine industries (Hiner, 2015), craft breweries are small businesses working with limited resources trying to make a name for themselves and build a consumer base.

While not unlike other small businesses, the circumstances of breweries—situated in cities, close to consumers, because freshness is a key component to quality beer—position them as key players in "buy local" campaigns. In this way, even if the ingredients of beer are coming from different, sometimes distant locales, the product, once produced, has a decided shelf life. As such, freshness in the product and local consumption are entwined. Moreover, brewers suggest that the "localness" of a given brew is created by the craftsman, the person who designs and creates a particular product in place with the materials—whether imported or not—at hand (Brewer, West Sacramento; 23 June 2015).

Beer is thus positioned to fulfill several transformative roles; beer/brewing serves as a vehicle of material transformation whereby rural inputs are transformed into an urban product and, related, globally sourced materials are converted into "local" ones. For example, water, hops, grain, yeast, which are often sourced from rural (and sometimes far-flung) locales, and yet these rural goods become a local (and often urban) product, reflecting a rural/urban metabolism (McKinnon, Hurley, Hiner, & Maccarroni, in revision). Related is how brewers take raw materials from all of the globe and transform them into a "local" product. As noted, one brewer suggested this transformation could be explained as the brewer is crafting a product in place; that is, the skills, knowledge, and context of that brewer is rooted in place and, thus, the resulting product is also placed—and therefore "local" despite the origin of some of its key inputs. Moreover, the short (comparatively to other products) shelf life of beer forces its perception as a "fresh" (read: "local") product. In this way, brewers and breweries are in essence creating "localness". In addition, by building up and participating in "buy local" campaigns and movements, breweries become related not only to processes of (neo) localism but also notions and motions of "sustainability" in the craft beverage industry.

In sum, the production of food, and its associated tourism, can play an important role in place and identity formation (Bell & Valentine, 1997; De Solier, 2013), and breweries, wineries, cider houses, and other fermented beverages, as a form of place-based food production, also help to foment and build local and regional identities (Hiner, 2015). In the context of these cases, collective tourism and marketing schemes create (micro) movements of change; whether transforming materials like raw, rural inputs into urban, value-added ones or transforming "forgotten" neighborhoods into desirable ones, beer serves as a catalyst for the (re) evolution of places. The possibilities for further research in this area are numerous and could include explorations of the specific economic impacts of brew trails or deeper investigations into the cultural contributions and/or environmental implications of such trails in the places in which they emerge and develop.

References

Alonso, A. D. (2011). Opportunities and challenges in the development of microbrewing and beer tourism: A preliminary study from Alabama. *Tourism Planning and Development, 8*, 415–431.

Bell, D., & Valentine, G. (1997). *Consuming geographies: We are where we eat.* New York, NY: Routledge.

Bizinelli, C., Manosso, F. C., Gândara, J., & Valduga, V. (2013). Beer tourism experiences in Curitiba, PR. *Rosa dos Ventos, 5*, 349–375.

Brewer Association. (2014). Craft brewer definition. Retrieved July 8, 2016, from https://www.brewersassociation.org/brewers-association/craft-brewer-definition/

De Solier, I. (2013). *Food and the self: Consumption, production and material culture*. New York, NY: Bloomsbury Academic.

de Wit, C. W. (2013). Interviewing for sense of place. *Journal of Cultural Geography, 30*(1), 120–144.

Farm-to-Fork Program, S. (2016). *The program: Welcome to America's farm-to-fork capital*. Retrieved July 1, 2016, from http://www.farmtofork.com/what-we-do/the-program/

Fastigi, M., & Cavanaugh, J. R. (2017). Turning passion into profession: A history of craft beer in Italy. *Gastronomica: The Journal of Critical Food Studies, 17*(2), 39–50.

Hall, C. M. (1999). Rethinking collaboration and partnership: A public policy perspective. *Journal of Sustainable Tourism, 7*(3–4), 274–289. doi:10.1080/09669589908667340

Herz, J. (2016). The importance of defining small and independent. Retrieved July 8, 2016, from https://www.brewersassociation.org/communicating-craft/importance-defining-small-independent/

Hiner, C. C. (2015). *Making (a) place: Wine and the production and consumption of landscapes in the Sierra Nevada foothills of California*. Quadrennial Conference of British, Canadian, and American Rural Geographers. Swansea & Aberystwyth, Wales, UK: University of Wales.

Hoalst-Pullen, N., Patterson, M. W., Mattord, R. A., & Vest, M. D. (2014). Sustainability trends in the regional craft beer industry. In M. Patterson & N. Hoalst-Pullen (Eds.), *Geography of beer* (pp. 109–118). New York, NY: Springer.

Holtkamp, C., Shelton, T., Daly, G., Hiner, C. C., & Hagelman, R. (2016). Assessing neolocalism in microbreweries. *Papers in Applied Geography, 2*, 66–78.

Larsen, S., & Hutton, C. (2011). Community discourse and the emerging amenity landscapes of the rural American West. *GeoJournal*, 1–15.

Mathews, V., & Picton, R. M. (2014). Intoxifying gentrification: Brew pubs and the geography of post-industrial heritage. *Urban Geography, 35*(3), 337–356.

McKinnon, I., Hurley, P. T., Hiner, C. C., & Maccarroni, M. (In revision). Uneven urban metabolisms: Toward an integrative (ex)urban political ecology of sustainability in and around the city. Urban Geography

Mittag, R. (2014). Geographic appellations of beer. In M. Patterson & N. Hoalst-Pullen (Eds.), *Geography of beer* (pp. 67–76). New York, NY: Springer.

Neister, J. (2008). *Beer, tourism and regional identity: Relationships between beer and tourism in Yorkshire, England*. Unpublished master's thesis, The University of Waterloo, Ontario, Canada.

Patterson, M., & Hoalst-Pullen, N. (2014a). *The geography of veer: Regions, environment, and societies*. New York, NY: Springer.

Patterson, M. W., & Hoalst-Pullen, N. (2014b). Geographies of beer. In M. Patterson & N. Hoalst-Pullen (Eds.), *Geography of beer* (pp. 1–8). New York, NY: Springer.

Plummer, R., Telfer, D., & Hashimoto, A. (2006). The rise and fall of the Waterloo-Wellington Ale Trail: A study of collaboration within the tourism industry. *Current Issues in Tourism, 9*(3), 191–205.

Plummer, R., Telfer, D., Hashimoto, A., & Summers, R. (2005). Beer tourism in Canada along the Waterloo–Wellington Ale Trail. *Tourism Management, 26*(3), 447–458.

Robertson, B. A. (2016). Map tells story of Sacramento's brewery growth. *The Sacramento Bee*, March 25.

Rogerson, C. (2015). Developing beer tourism in South Africa: International perspectives. *African Journal of Hospitality, Tourism and Leisure, 4*(1), 1–15.

Schnell, S. M., & Reese, J. F. (2003). Microbreweries as tools of local identity. *Journal of Cultural Geography, 21*(1), 45–69.

Schnell, S. M., & Reese, J. F. (2014). Microbreweries, place, and identity in the United States. In M. Patterson & N. Hoalst-Pullen (Eds.), *Geography of beer* (pp. 167–188). New York, NY: Springer.

Spriggs, B. (2014). Lexington downtowners on the G word: Gentrification. *Ace Weekly*, June 30.

Trubek, A. B. (2008). *The taste of place: A cultural journey into terroir*. Berkeley: University of California Press.

Yool, S., & Comrie, A. (2014). A taste of place: Environmental geographies of the classic beer styles. In M. Patterson & N. Hoalst-Pullen (Eds.), *Geography of beer* (pp. 99–108). New York, NY: Springer.

CHAPTER 11

Brewing a Beer Industry in Asheville, North Carolina

Scott D. Hayward and David Battle

INTRODUCTION

Beer tourism has taken off in Asheville, North Carolina. With 18 brewers and another 40 in the surrounding area, beer has become a key tourist attraction as have the city's many beer tours/tastings, beer festivals, and beer conferences. The concentration of brewers and attendant activities as well as a "supportive community with local pride" (Myers, 2012, p. 59) have made Asheville "a mecca for beer enthusiasts" (Krug, 2010, n.p.) and a frequent winner of the title Beer City, USA.

The brewery phenomenon is relatively new to Asheville. In 1994, Oscar Wong—an engineer by trade—moved to Asheville. He established Highland Brewing in the basement of a pizzeria with a vision to create a local beer similar to those in the small towns of Europe. Wong's success spurred growth in the number of area brewers, recently culminating

S.D. Hayward (✉)
Department of Management, Elon University,
Elon, NC, USA

D. Battle
Clover, SC, USA

with Sierra Nevada and New Belgium breweries locating their East Coast operations in Asheville.

Although local industry growth is an established vein of academic study, little theory addresses the development of regional identity and tourism's role in that process. In this chapter, we use interviews and secondary sources to document the rise of the craft beer industry in Asheville. We highlight the importance of creating opportunities for the interactions that drive a local industry's identity development. Early entrepreneurs linked a national trend to local resources creating a catalyst for subsequent growth. A core of companies and individuals cultivated social ties and institutions that, we propose, allowed people within the industry to interact and construct the local industry's identity.

While social interactions among local industry insiders are critical for identity development, we propose that local industry insiders also engage with residents, expressing and reflecting an identity of their city as a good place for craft brewing. Furthermore, we suggest that local industry insiders also express and reflect with tourists and other outsiders to further develop local industry image. Typically, research views image expression as the domain of tourism officials and government agents; its importance has been well documented in the literature (Gallarza, Saura, & Garcia, 2002). We propose a feedback effect where the expressed image is received by tourists and reflected back to the local industry reinforcing the local identity, or causing the local industry to revise the identity they hold or the image they express.

While the local industry's interactions with both residents and tourists shape the local identity, we consider that the interactions between residents and tourists likely matter as well. Here, we build on previous theory of authenticity, proposing that the authentic experience discussed in the tourism literature is a component of the local identity process. Through their interactions, residents and tourists express and reflect upon the local industry identity they hold. We propose that when the tourist and resident images agree, it confirms the identity held by the tourists establishing the experience's authenticity.

This chapter begins with an overview of personal interactions as an essential mechanism for identity development. We then describe the national and local conditions that preceded the establishment of the Asheville craft beer industry. In describing the development of Asheville's craft beer industry, we propose a triad of interactions that lead to local industry identity and the importance of creating space and opportunities where those interactions can occur.

Regional Identity as a Relational Construct

Our conception of local identity draws to the local level as an understanding of identity from organizational and social psychological research. Local identity consists of the perceptions, meanings, and knowledge shared among people about a place. For residents, local identity describes "who we are"—what is central and unique about us—as a locale. For non-resident outsiders, the locale's identity are the perceptions and meanings they associate with the place (Albert & Whetten, 1985; Romanelli & Khessina, 2005; Wang & Chen, 2015). While this definition is consistent with other studies of regional identity in the tourism literature (e.g., Wang & Chen, 2015), we deepen our understanding of local identity when we consider the underlying micro mechanisms that form this identity. In particular, it behooves us to recognize identity, including place identity, as a product of interactions and discourse between people and groups. As such, the place identity individuals hold in their minds, and how they describe it, may differ. Following the organizational theory and social psychology literature, throughout this chapter, we separate the concepts of identity and image: while identity is that which is held by individuals or groups, an image is the identity expressed by those entities to others.

A relational approach to identity notes how identity forms as people interact with one another. As individuals interact with others, they compare their notions (of themselves, their organizations, their locales, etc.) with those held by others, making adaptations to their beliefs and adjusting their perceptions accordingly. At the local level, we propose that local identity forms from two processes, both of which depend on social interactions. First, each incorporates her perceptions in her discourse with others. People make claims about where they live and the places they know as a form of self-expression and a way to communicate what they think is central and defining about a location and its people. Individuals present their area's image through a process of expression. Second, after expressing the image, individuals view and consider the reactions of and images expressed by others through a process of reflection. When the images received by an individual complement the identity held by that individual, reflection reinforces that identity. When the reflection contradicts the identity an individual may hold, she updates either how she expresses that image or adapts her held identity accordingly (Hatch & Schultz, 2002).

The implications of this relational view of local identity are twofold. First, to understand local identity, we must understand how individual perceptions evolve into the images regions project to outsiders. Second, given the importance of interactions, we must understand how a local industry can create the opportunities for interactions to occur. We find the emergence of the Asheville craft beer industry an ideal case for theorizing on the potential mechanisms in play for local identity development. We provide a short history of the rise of the industry and explore the relational view of identity as part of that process. Given the experiential nature of craft beer and the craft brewery business model, this locale and industry provide an ideal case for theory development.

SETTING THE STAGE: NATIONAL MODELS AND LOCAL RESOURCES

Local industries and identities do not form from nothing. The nascence of Asheville's craft beer industry occurred as entrepreneurs accumulated and combined local resources with a nationally established business model, to seed new ventures in the area.

US Craft Beer Identity in the 1990s

In the mid-1990s, the US craft beer industry hit its stride. From 18 microbreweries a decade earlier, the industry grew to 537 craft breweries with considerable room for more (Hindy, 2014). Most began as brewpubs (a restaurant brewing its beers), then some brewpubs expanded into packaging and distributing their beer more widely (Hindy, 2014). Government regulations and downstream oligopolies, however, made broader distribution a considerable challenge (Baron, 2009).

At the same time, investment and national beer companies began paying attention to the public's interest in craft brews. Capital markets, for one, took a number of craft brew pioneers public (e.g., Boston Beer Company, Redhook Ale). Industry giants like Anheuser-Busch invested in microbreweries, imitated craft beers, and pressured craft brewers through various tactics including attempts to limit their distribution. Despite these manipulations, there was safety in being small and new craft breweries continued to open (Hindy, 2014).

Craft brewers juxtaposed themselves with the large "industrial" brewers and imports by defining themselves as "chefs" and "lunatics at the fringe". They focused on their creative, artistic quality craftsmanship, while incumbent brewers seemingly focused on maximizing shareholder's wealth (Baron, 2009; Hayward & Jiang, 2016). As one brewer proclaimed in a trade association magazine: "Our recipes, like prose, exist on paper though ultimately live and breathe in our customer's glass… our recipes are not merely words on a page. They are pieces of liquid art which evolve over time and provide thoughtful discussion" (Arthur, 2000, quoted in Hindy, 2014, p. 115). Thus, the craft beer market not only developed in terms of sales, financing, and production, it developed in terms of the identity expected of the craft brewing business. Lamertz, Heugens, and Calmet (2005) found that identities about the brewing process split from scientific control and economies of scale on the one hand, and craftsmanship and specialty on the other. Craft brewers are often defined in juxtaposition with the big brewers. Furthermore, Lamertz et al. (2005) found that craft brewers often identified the authenticity and expressive values of their beer and their close affiliation with a local market and community. The value to craft brewers in identifying themselves by their small size and local connections is well documented (Flack, 1997; Lamertz, Foster, Coraiola, & Kroezen, 2016; Schnell & Reesse, 2003), and invoke the phrase "neolocalism" to capture the "conscious creation and maintenance of attachment to place" (Flack, 1997, p. 45). Neolocalism is broader than just craft beer, nationally witnessed by movements for local foods, local business support, and local identity (e.g., "Keep Austin Weird"). Yet, in craft beer, neolocalism manifests itself in the names of local breweries, names of local beers, and the decor and labels the breweries and beers display (Schnell & Reesse, 2003).

Asheville's Identity Resources

The craft brewery model and its artisan identity spread nationwide before settling in Asheville. Moreover, the identity model already carried with it some expectations of building upon local culture and catering to the local market. Two aspects of Asheville's history created conditions in which the craft breweries could thrive. First, Asheville had a history of tourism, cultural tourism in particular. Located in the southern Appalachian Mountains, Asheville's pleasant climate and natural beauty had historically attracted wealthy elites looking to escape and to convalesce from tuberculosis. Post–Civil War industrialists like Cornelius Vanderbilt built local estates and financed local improvements to make Asheville a "playground for the wealthy" (Starnes, 2005, p. 51). These improvements included many art-deco style buildings that gave Asheville a distinct architectural feel. Furthermore, these wealthy tourists encouraged and purchased local crafts and music which provided "a springboard for profits and cultural preservation" (Starnes, 2005, p. 183). Hence, for 100 years before its first, modern microbrewery, local culture was an integral part of Asheville's tourism industry, giving Asheville a strong sense of place both at home and abroad.

Despite the efforts of Vanderbilt and others, tourism could not save Asheville from the downturns of the 1930s and 1970s. Businesses left many downtown buildings vacant after moving to the city's suburbs. In the mid-1980s, Asheville adopted a public-private development model credited with much of Asheville's current success. Rather than raze derelict buildings for large-scale projects, stakeholders preserved local buildings, rebuilt civic amenities, encouraged downtown residency, and supported the growth of local, independent businesses. City government, with community backing, established development plans and institutions to preserve Asheville's downtown. For example, city government initiated the Asheville Downtown Association (ADA) to represent downtown business interests. While initially funded and staffed by the city, it is now an independent non-profit entity. Private investment was also involved in Asheville's success. Julian Price, a wealthy insurance heir, established the Public Interest Project (PIP). PIP invested in residential units and funded many small businesses (Strom & Kerstein, 2015b).

Importantly, the timing of Asheville's redevelopment coincided with the rise of the national microbrewery industry. Asheville's public (e.g., ADA) and private (e.g., PIP) organizations benefited local craft brewers directly (the ADA now organizes downtown events highlighting Asheville's microbreweries) and indirectly (PIP-funded companies encouraged subsequent related businesses or even started their own microbreweries) (Finlay, 1994). The nature of Asheville's redevelopment fit the needs of craft brewery entrepreneurs. A decade of downtown redevelopment made available an interesting built environment and a supportive entrepreneurial culture ripe for businesses like craft breweries. Moreover, Asheville's identity as a community supportive of local, creative businesses fit the identity of typical craft brewers. One brewer we interviewed suggested that neolocalism, and the strength of residents' attachment to the place, was particularly strong in Asheville. When asked why the craft beer market grew, the founder described local attachment as having the largest effect on the growth of his business:

> *I think Asheville is a lot more locally focused. I would say that if we started this in Charlotte, it wouldn't have gotten so much local support. The size of this place seems to indicate people know about things a little more because it is not that big of a town. And they support local....Localism has strong roots in Asheville's mountain culture, as witnessed by the thriving Farm to Table and the Slow Food movements.*

REGIONAL INDUSTRY DEVELOPMENT

Combined with social identity theory, the story of Asheville's craft beer industry suggests an underlying process by which a local industry develops an identity in stages of interactions. Entrepreneurs provide the catalyst for the industry's development, drawing resources from home and abroad to give local flavor to a national trend. Early insiders interact with each other, carrying their perceptions about not only their companies but about the region and its influence on the local industry. Local industry identity, we suggest, is born through the networks and institutions that facilitate insider interactions.

Local industry identity evolves, we propose, through the interactions industry insiders have with residents and outsiders (i.e., tourists and distant consumers). The development of Asheville's craft beer industry created the infrastructure where these types of interactions could take place. Through these three types of interactions: intra-industry, industry-resident, and industry-outsider, Asheville's craft beer identity formed and matured. The following sections relay Asheville's history, as we highlight some of the key mechanisms it suggests.

Phase One: Entrepreneurship

While scholars have theorized about why industries expand in a particular location, how those local industries start is attributed to fortuitous circumstance: "[E]arly firms are put down by historical accident in one or two locations; others are attracted by their presence, and others in turn by their presence. Then industry ends up clustered in the early-chosen places" (Arthur, 1994, p. 50). For Asheville's beer industry, the "historical accident" appears to be a vacation home. In 1994, a retired engineer named Oscar Wong needed something to do with his time. He turned his hobby of home brewing into a local beer similar to those in the small towns of Europe. He did this by teaming up with John McDermott, a brewer at a Charlotte, North Carolina, brewery, and stipulated that they must build the business in Asheville where he had a second home. Barley's Taproom, a pizzeria with beers on tap, was opening in downtown Asheville and offered Wong space for his brewery in the basement. There, the two entrepreneurs founded Highland Brewing Company (HBC) with used dairy equipment and a steep learning curve: Wong famously dumped 3000 gallons of Highland's earliest beer because it did not meet his quality standards (Glenn, 2012).

While Wong wanted to focus on brewing, not the restaurant side of a brewpub, Barley's provided an immediate outlet for retail sales. The model proved lucrative. The company built its brand through a loyal following and hand-filled 22-ounce bottles that moved it beyond the local draught market (Kiss, 1995). Three years after the introduction of HBC, another Asheville brewery opened. The owners of an established, PIP-funded Asheville restaurant hired Jonas Rembert, a brewer from Atlanta, to start Green Man Brewery. His mission was to brew beer to sell at the restaurant's pub. As the owner explained: "Everything I've done, I've done from necessity. You need fresh bread, you make it. You need fresh beer, you make it" (Glenn, 2012, p. 75). Unlike HBC, Green Man Brewery began with the idea that they would sell the product at a brewpub targeting the downtown clientele.

One year later, a third brewery—Asheville Brewing Company (ABC)—began in an old movie theater. The owners, a local restaurateur and a relocated brewer, offered beer, pizza, and second-run films. The relocated brewer, Doug Riley, had gone to college in Asheville but moved to Portland, Oregon, to learn to brew. In the early years, beer was more of a talking point than a profit center. Riley catalyzed a following with quality beer, but with low volume and other craft brews on tap. He treated the brewpub portion of ABC as support for pizza sales.

While the entrepreneurs reached out of Asheville for critical resources and knowledge, they leveraged local identity resources. Two of the three earliest breweries leveraged place in the brewery name ("Highland" and "Asheville"). All three earliest breweries located in interesting, downtown spaces which fit with the broader craft beer identity.

Proposition 1 Local industrial identity begins with entrepreneurs combining non-local models with local identity resources.

Phase Two: Intra-industry Interactions

With the establishment of three microbreweries, the growth of Asheville's craft beer industry gained momentum (Table 11.1). First, the three microbreweries created a pool of trained workers who would soon become the founders of future craft breweries. For example, in 2001, Jonas Rembert used his business and brewing experience from Green Man Brewing to cofound French Broad Brewing Company (Kiss, 2001). Notably, French Broad Brewing Company carved a new niche for itself by opening a small tasting room for consumers. According to Buenstorf and Klepper (2009), regional growth often comes from young, small firms who spawn new startups, which typically remain local.

As Fig. 11.1 illustrates, during this period of endogenous growth, craft breweries in Asheville became a training ground for future brew masters. Like the national trend, the Asheville craft beer industry grew more from new entrants than from the growth of existing craft breweries, despite the potential cost savings associated with being big (e.g., McGahan, 1991).

While these brewers took with them the knowledge required to start an Asheville brewery, they also created personal and professional social ties which crossed organization boundaries. These social relationships provide paths for identity-shaping interactions.

Table 11.1 Asheville craft brewery growth timeline (2009–2012)

Entrant	Brewer	Established	Location
1	Highland Brewing Co.	1994	Downtown
2	Jack of the Wood Green Man	1997	Downtown
3	Asheville Pizza and Brewing	1998	North Asheville
4	French Broad	2001	South Asheville
5	Pisgah Brewing	2005	Black Mountain (East)
6	The Wedge	2008	River Arts District
7	Craggie Brewing	2009	Downtown
8	Oyster House Brewing	2009	Downtown
9	Lexington Ave. Brewery	2010	Downtown
10	Altamont	2012	West Asheville
11	Wicked Weed	2012	Downtown

180 S.D. HAYWARD AND D. BATTLE

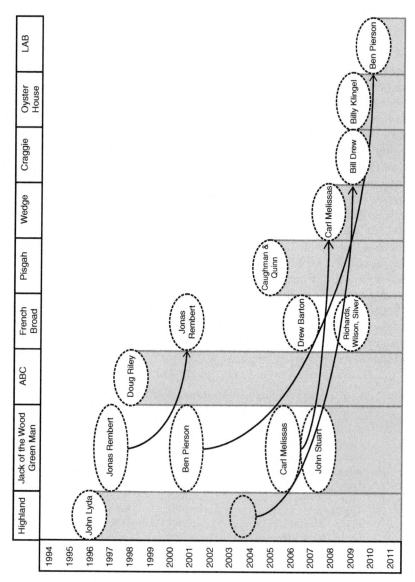

Fig. 11.1 Brewmaster movement, 1994–2011

Proposition 2a Local industrial identity begins with local intra-industry interactions facilitated by social ties.

Identity defining interactions also occur through problem-solving and conflict resolutions (Jones, 2011). Relationships between breweries in Asheville, which had long been pleasant and helpful (e.g., Glenn, 2012, p. 81), were formalized in 2009 after a disagreement between two firms over a beer name. Industry participants formed the Asheville Brewers Alliance (ABA) with a mission to benefit the collective (Asheville Brewers Alliance, 2016). The ABA facilitates communication and resolves conflicts among regional craft breweries and works together for mutual opportunities. Brewers clearly understood the institution's role in creating a unified image. One brewer noted the need to unite, lobby, and influence legislation. Other brewers had tourism in mind: "The brewery market was starting to get crowded. We weren't Beer City yet, but we could see beer tourism on the horizon. And we knew that it would be good for all of us" (Glenn, 2012, p. 82).

Proposition 2b Regional industrial identity evolves through local intra-industry interactions facilitated by local industry institutions.

Phase Third: Extra-industry Interactions

The perceptions outsiders' hold of a location's unique capacity for a particular industry further encourages support (Romanelli & Khessina, 2005). Asheville's experience suggests a third phase of identity development where the local industry expresses its identity to three sets of customers: residents living in the location, tourists, and distant consumers. While remote consumers are audiences for the local industry's image, we propose that residents and tourists have a direct impact on identity development. We propose the following triadic model of interactions to better understand local industry identity dynamics, and the role played by tourists (Fig. 11.2).

Media as Image Expression: The Kiss Effect

Local media disseminate information and, perhaps more importantly, profile people and events in particular ways (Anderson, 2006; Romanelli & Khessina, 2005). Asheville's media, and a writer for its largest local paper,

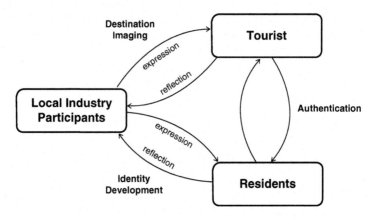

Fig. 11.2 Triadic model of local identity interactions

in particular, was one important driver of identity construction. Tony Kiss, the Entertainment Editor for the Asheville Citizen-Times, reported on the emerging craft beer industry from its inception in the basement of Barley's. As the industry grew, so did Tony Kiss' coverage and encouragement, further legitimizing the industry to his readers.

Tony Kiss not only shaped the identity held by Asheville's residents, but he also helped express Asheville's image with a non-resident audience. In 2009, a renowned craft brewer and founder of the National Brewers Association held an online poll to crown the country's best beer city, "Beer City, USA". Driven by Kiss' strong promotional push, Asheville tied Portland, Oregon, that year and won or tied for the title the next three years. In a 2010 interview, Kiss noted the pride the local craft brew industry took in the title and how it had "added to the overall excitement [of] the brewery scene here". Externally, the title "put us even more on the beer map. It's given us national attention, and let people…know about Asheville" (Kiss, 2010, n.p.). While the Brewer's Association poll lacked scientific rigor, the title, "Beer City, USA" clearly reinforced the Asheville's craft brew industry's identity for industry insiders, residents, and outsiders alike.

Proposition 3a Local industrial identity evolves through reflection generated by local media.

Experiencing the Industry: Creating Space for Interactions with Residents and Tourists

Asheville's experience indicates that tourism played a critical role in developing the local craft beer identity. The growth of Asheville's beer industry coincided with a rejuvenation of Asheville's tourism industry. In 1985, city and county leaders pushed through a hotel occupancy tax that was used to finance advertising for the area. As a result, Asheville's tourism industry grew in both annual visits and amenities. Between 1990 and 2013, Asheville's tourism revenue rose 230 % (Strom & Kerstein, 2015a). In 2014, the tourism industry was the area's third largest employer.

Advantageous to Asheville's craft beer identity (and perhaps resulting, in part, from the craft beer industry), tourism in Asheville remains more about niche than mass tourism (Strom & Kerstein, 2015a). Compared to the typical US tourist, visitors to Asheville are 30 % more likely to participate in culinary experiences, 40 % more likely to participate in festivals/fairs, 240 % more likely to participate in adventure sports, and 800 % more likely to participate in beverage tours (Kohler, 2014). Focusing on the specific needs of niche tourists may have helped Asheville build its image and avoid problems arising from mass tourism (e.g., private spaces, seasonal labor pools, degraded natural resources) (Strom & Kerstein, 2015a).

Similar to wineries, who initially did not regard their businesses as tourist attractions, Asheville's craft breweries did not focus on tourism until the mid-2000s and after Asheville was named Beer City, USA. Also like wineries, craft brewers responded to the potential of tourism by emphasizing the experience as well the product (Getz & Brown, 2006; Howley & Westering, 2008). Asheville's craft beer industry now offers three experiences that directly link their breweries to residents and tourists: tasting rooms, beer tours, and festivals. While it is a source of revenue and customer loyalty, they also provide the opportunity for interactions at the heart of local identity creation.

Tasting Rooms: Similar to the winery cellar door (Fraser & Alonso, 2006), the tasting room at craft breweries became a critical part of attracting loyal customers. There, brewers shaped consumers' experience while driving sales, promotion, and education. Having tasting rooms may be particularly important for craft brewers who have historically faced challenges in distribution (Baron, 2009; Dunn & Kregor, 2014). In 2004, French Broad Brewing Company opened with a small 25-seat tasting room, which offered consumers the space to taste beers and listen to live music in an intimate setting, making it an essential part of the business (Myers, 2012). Alternatively, Pisgah Brewing Company, located in nearby Black Mountain, North Carolina, only offered tastings in what was formerly a furniture manufacturing plant on Thursday afternoons. The popularity of the Thursday tastings encouraged the founders to open a taproom with multiple taps and a music stage. They also cleaned up a nearby meadow where customers could congregate. Five years later (i.e., 2010), they added a large, outdoor stage to the meadow, which has since become a music venue for national acts (Glenn, 2012). "The music just came about", one founder recalled. "I'm a music lover, so we just kind of went in that direction. We built a stage instead of buying more tanks" (Myers, 2012, p. 98). By 2012, all of Asheville's craft brewers had a restaurant and pub or taproom to reach consumers directly. When done correctly, suggests McBoyle and McBoyle (2008), reaching consumers directly creates an emotional connection between the consumer and the craft brewery that drives brand loyalty.

Beer Tours: While each craft brewery individually creates a consumer experience with their taproom, outside entrepreneurs have established beer tours that link these taprooms for an integrated customer experience. Research suggests that beverage producers hesitate to enter the tourism market. In part, this may be due to the challenges of working with tourism organizations and the mass tourists they target (Hall et al., 2009). It may also be due to business' hesitation to cooperate with competitors (Martin & Haugh, 1999). In Asheville, it was local entrepreneurs rather than a brewer who eventually linked the brewery experiences through a tour.

Encouraged by the success of Pisgah's taproom and the media coverage by Tony Kiss, in 2006, Mark and Trish Lyons built a tour company focused on visits to multiple microbreweries. As Mark Lyons recalled, breweries were willing to give our "Asheville Brews Cruise" a try; they offered tours of their facilities, educated tourists in the production process, and provided samples and parting gifts (Lyons, n.d.). The success of the Asheville Brews Cruise sparked an entrepreneur with experience in food tours (2010) and a former bartender (2013) to offer similar tours. Illustrating the popularity of the beer tour, the Lyons' have since franchised the Asheville Brews Cruise model to other cities.

Beer Festivals: Beer festivals also contributed to the growth of Asheville's craft beer industry. A growing body of literature highlights organizational motivations to generate events that attract tourists and provide an economic benefit (Stokes, 2008). Event agencies, tourism marketing authorities, event directors, and city and local governments play a role in supplying and marketing such events. Each organization tends to exist within a network of organizations with goals, ideas, and reasons for creating an event. Locations differ in the potential they have for growing such tourism (Prentice & Andersen, 2003). Thus, to understand the growth of beer festivals, it is important to understand the drivers of supply and demand in Asheville. Local business associations, local beer retailers, local beer-tourism firms, and the brewers themselves organize craft beer events in Asheville. Each has specific motives and resources, yet collectively contributes to Asheville's overall identity. Interestingly, the City of Asheville does not organize events but is certainly an integral part of the process by providing space, permits, and promotion.

In 1997, Asheville held its first craft beer festival. Three years after the establishment of HBC, the founders of Barley's Taproom began the Great Smokies Craft Brewers Brewgrass Festival (a.k.a. "Brewgrass"). At the time, only HBC and Green Man Brewery were brewing craft beer in Asheville, and the sale of local beers was not their primary motive. Barley's specialized in draft beer, but with limited demand, the owners recognized that they needed to educate consumers and introduce them to a wider variety of beers. Brewgrass's initial mission was to combine craft beer with other aspects of Asheville's identity and culture. Said one founder: "We wanted something that would make us unique among beer festivals and showcase what Asheville is all about…We want to turn this into a major bluegrass festival, not just a beer event" (McGee, 1998, n.p.). Brewgrass's fortunes grew along with the growth of the local craft beer industry, becoming one of the Southeast's premier festivals, but struggling to maintain its "boutique" feel while experiencing increased demand.

Other events link Asheville's craft beer to other aspects of Asheville's culture while providing space for interactions among the industry, residents, and tourists. The Winter Warmer Beer Festival, first held in 2008, was organized by the owners of the Asheville Brews Cruise. Held indoors and during the winter, the festival is small but links local craft beers to local restaurants and local music. Additionally, the Ashville Downtown Association (ADA) organizes Oktoberfest to attract people to downtown businesses. While the sale of local craft beers was not the primary goal of the festival, the ADA expanded its initial plans to include opportunities to meet local brewers and partake in beer dinners (highlighting beer and food pairings). Asheville also has two home brew competitions, one hosted by a local homebrewing club (begun in 1999) and the other a fundraiser for a local progressive organization (launched in 2010) (Douglas & Arnaudin, 2015).

The latest event to be offered is put on by the consolidated local industry itself. After winning the title "Beer City, USA" for the second time, the ABA (in conjunction with the Brewgrass Festival organizers) began the "Beer City Festival". This festival has a local, educational purpose with tickets sold only in Asheville. In 2012, this festival became part of the Asheville Beer Week. Local beer retailers, distributors, and brewers founded AVL Beer Week to "[facilitate] a cooperative atmosphere among the drinking community, local breweries and other Asheville businesses". Organized by the ABA, AVL Beer Week includes dinners, brewery tours, and the Beer City and Just Brew It Homebrew Festivals (Penland, 2016, n.p.).

The establishment and development of Asheville beer festivals were critical to the local industry's growth. They draw in both residents and tourists; organizers sold half of the Brewgrass's 3500 tickets outside of Asheville. The festivals also link the local craft beer industry with other local industries. For example, in its most current format, AVL Beer Week educates and socializes individuals and draws in other businesses:

> The goal was to shine a light on the really cool stuff that was happening here in the craft brewing industry, but also to help invigorate the community and give an opportunity for other businesses to benefit by bringing more folks here. Businesses don't even have to serve alcohol to participate or host an event. It's been interesting for us to be able to partner and collaborate with other local businesses. (Glenn, 2010, n.p.)

As Asheville's craft beer industry offered experiences to residents and tourists, it created interactions for expressing and reflecting Asheville's craft beer image.

Proposition 3b Local industrial identity evolves through expressing and reflections between the industry and residents/tourist that occur through shared experiences.

Extending this line of thought further, we highlight how these events create a sense of place for both tourists and residents (De Bres & David, 2001; Derrett, 2003). Balancing both residents' and tourists' needs is not unique to Asheville's craft beer industry, but reflects Asheville's broader approach to development. The revitalization of Asheville's downtown and other neighborhoods provides space to be shared by both locals and tourists. While local officials encourage tourism, they also protect residents as exemplified by the City Council's recent decision to prohibit the Brewgrass Festival from its longtime location in a local park (Asheville City Council, 2013). Furthermore, Asheville hosts a multitude of creative industries beyond craft beer. A mountain arts and crafts industry and a regional cuisine industry thrive in Asheville. They link tourists and residents to Asheville's identity in much the same way as the craft beer industry through local studios and restaurants, "trails" and "hops", and festivals, with lots of crossover among them.

While cities strive to attract creative residents and tourists, it may be the interaction of the two that provides Asheville with the symbolic edge it needs to compete. With its taprooms, tours, and festivals, Asheville's craft beer industry has created space for residents and tourists to interact and share an experience, shaping "authenticity", and avoiding the "commoditization" that often is a result of tourism development (Strom & Kerstein, 2015a). Tourists often seek an experience that honestly reflects local culture. As such, they look for cultural products and experiences that portray to visitors the same meanings that are accepted by locals. Commoditization arrives when officials stage cultural experiences for the benefit of tourists (MacCannell, 1973; Ram, Bjork, & Weidenfeld, 2016).

A relational view of identity suggests that for a tourist's experience to be authentic, it must fulfill two criteria. First, the local identity held by tourists and the image expressed by residents must match in "content and emotional tenor" (Howard-Grenville, Metzger, & Meyer, 2013, p. 115). For Asheville's tourists, an experience that matches the identity must hold with the image conveyed by residents and reinforce that identity. Attending a festival that highlights a large number of local breweries reinforces that city's image as a "good beer town".

An authentic experience breaks down the difference between residents and tourists, with both attaching similar meanings to the same event. Further, images and identities are authenticated when residents and tourists observe and interact with each other in ways that allow them to compare how each experiences the event. Taprooms, tours, and beer festivals create such shared experiences where residents and locals are also consumers. As the following model illustrates, tourists interacting with residents likely compare their image with the resident's identity. A resident's local image which matches a tourist's local identity reinforces the tourist's local identity and proves the experience authentic. Tourism boards and local officials construct an image for tourists, and tourists put that image to the test during their experiences.

Extending Strom and Kerstein's (2015b) argument, we suggest that the experiences offered by Asheville's craft beer industry experiences have catered to residents and tourists alike, allowing both to interact in the "front stage" and compare their image and identity. Residents too may be affected by their interactions with tourists. Congruent with organizational theory (e.g., Dutton & Dukerich, 1991), we suggest that the feedback from tourists also shape how Asheville's residents feel about (and act within) their city. With the convergence of image and identity, the experience has been authenticated, benefitting Asheville, its residents, and tourists who recognize the unique experience that is offered by Beer City, USA!

Proposition 4a An authentic experience for tourists requires the opportunity for tourists to interact with residents.

Proposition 4b An authentic experience for tourists results from complementary identities held by tourists and residents.

Epilogue: Exogenous Entrance as Identity Validation

Maturing local industries also attract industry incumbents and investments from outside the region. While researchers often cite externalities and natural resources as the reason for growth, a healthy, local image likely attracts resources through tourism, migration, and investments. External audiences understand key features of work and life in the region and can assess the benefits of locating there (Romanelli & Khessina, 2005). Indeed, by 2012, Asheville had captured the attention of existing craft brewers searching for a brewery-distribution location which would reduce transportation costs to eastern US markets. Two of the largest US craft brewers, Sierra Nevada and New Belgium, turned down competing economic incentives from other municipalities to leverage Asheville's identity and resources (Ferguson, 2012). Sierra Nevada cited water quality, quality of life, outdoor culture, and craft beer culture as its reasons for locating in Asheville (Ferguson, 2012). New Belgium recognized Asheville's commitment to bike paths, greenways, and alternative transportation in conjunction with the continued support of the River Arts District (Glenn, 2012).

Some local brewers met the announcement of Sierra Nevada and New Belgium move into Asheville with some trepidation. Yet, many brewers focused on the benefits of large and well-known players as confirmation of Asheville's craft beer identity and its ability to attract tourists. For their part, Sierra Nevada, New Belgium, and other new entrants have taken steps to become partners in the local industry and build facilities and events that promote tourism. In doing so, they not only reflect the local industry identity back to local craft brewers but also create the space to further its identity development and authenticity.

Conclusion

Asheville's history and identity created a natural terroir for a thriving craft beer industry. Established as a cultural-tourism destination, its redevelopment in the 1990s coincided with the growth of craft beer nationwide. Asheville's renewal resulted in shared spaces for tourists, residents, and craft brewers as well as a downtown that attracts creative workers (Scherer, 2007) and weekend visitors from multiple Southeastern cities. Both factors have allowed Asheville to support more breweries per capita than any other place in the country (Baginski & Bell, 2011; Strom & Kerstein, 2015a).

The evolution of the Asheville craft beer industry provides a motivating case for understanding local industry identity dynamics. We began this chapter defining regional identity as a relational construct, noting the importance of interpersonal interactions in the identity construction process. Grounded in psychological and organizational theories (Albert & Whetten, 1985; Hatch & Schultz, 2002; Jones, 2011), we proposed a process-based theory of regional identity where the local industry expresses its image to residents and tourists and adjusts its identity and actions based on their response. We suggest that the many opportunities for the local industry insiders to interact with each other and with residents and tourists were crucial for Asheville's craft beer identity. Intra-industry social networks, local trade institutions, tasting rooms, tours, and festivals are the stages for identity development in craft beers. We conclude by considering resident-tourist interactions and defining authentic experience as the match between tourist and resident images of the local industry.

This chapter contributes to tourism studies by extending our understanding of identity as more than an exercise in marketing, but as a social process. Identity emerges from the interaction of key stakeholders, with tourism officials playing one role in expressing an image to outsiders. By articulating the mechanism linking stakeholders to identity, we highlight the need to study not only the message presented by tourism officials but also the business models that allow tourists to connect with the local industry and residents.

In a practical vein, knowing how local identity dynamics work helps local industries develop institutions and opportunities for interactions. In addition to coordinating among local industry participants, industries should nurture the triad of interactions set forth in this model. An identity balanced across these stakeholders are more resilient as people, firms, and conditions change. Rather than a one-way pitch from the industry to outsiders, local resiliency requires open conversations among the industry, residents, and tourists.

The case we have presented raises more questions than answers. We hope, however, that propositions from Asheville's craft breweries bring a new perspective to the literature. Tourism researchers have many methods available (cf. Gallarza et al., 2002). As we move forward, however, indepth cases employing interviews, archival documents, and press accounts may allow us to see tourism in an expanded light. Cities consist of interacting stakeholders each with their viewpoints and ambitions. They engage each other through negation, problem-solving, and interrelated actions. In this sense, tourists are more than customers to market to or sources of economic rents. Rather, they are distinctive stakeholders, playing their part to shape the local social fabric.

References

Albert, S., & Whetten, D. A. (1985). Organizational identity research. *Organizational Behavior, 7*, 263–295.

Alliance, A. B. (2016). Asheville Brewers Alliance.

Anderson, B. (2006). *Imagined communities: Reflections on the origin and spread of nationalism.* New York: Verso Books.

Arthur, T. (2000). Inventing style for a brave new world. *New Brewer Magazine.*

Arthur, W. B. (1994). *Industry location patterns and the importance of history increasing returns and path dependence in the economy.* Ann Arbor: The University of Michigan Press.

Asheville City Council. (2013). City council meeting minutes. Asheville, NC. Retrieved from http://www.ashevillenc.gov/Portals/0/city-documents/cityclerk/mayor_and_citycouncil/minutes/mmthirteen/m130813.pdf

Baginski, J., & Bell, T. L. (2011). Under-tapped?: An analysis of craft brewing in the Southern United States. *Southeastern Geographer, 51*(1), 165–185.

Baron, A., Person, D. (Producer), & Baron, A. (Director). (2009). *Beer Wars* (motion picture). USA: Gravitas Ventures.

Beebe, C., Haque, F., Jarvis, C., Kenney, M., & Patton, D. (2013). Identity creation and cluster construction: The case of the Paso Robles wine region. *Journal of Economic Geography, 13*(5), 711–740.

Brewers Association. (2015). *Craft beer sales by state.* Retrieved from https://www.brewersassociation.org/statistics/by-state/?state=PA

Buenstorf, G., & Klepper, S. (2009). Heritage and agglomeration: The Akron tyre cluster revisited. *The Economic Journal, 119*(537), 705–733.

De Bres, K., & David, J. (2001). Celebrating group and place identity: A case study of a new regional festival. *Tourism Geographies, 3*(3), 326–337.

Derrett, R. (2003). Making sense of how festivals demonstrate a community's sense of place. *Event Management, 8*(1), 49–58.

Douglas, S., & Arnaudin, E. (2015). Brews on the move: A look at Asheville' beer tour industry. *Mountain Xpress.* Retrieved from http://mountainx.com/food/local-brewery-tours-part-one/

Dunn, A., & Kregor, G. (2014). Making love in a canoe no longer?: Tourism and the emergence of the craft beer movement in California. CAUTHE 2014: Tourism and Hospitality in the Contemporary World: Trends, Changes and Complexity, 189–197.

Dutton, J. E., & Dukerich, J. M. (1991). Keeping an eye on the mirror: Image and identity in organizational adaptation. *Academy of Management Journal, 34*(3), 517–554.

Ferguson, M. L. (2012). *Why Sierra Nevada fell for Asheville.* Retrieved from http://therevivalist.info/sierra-nevada-asheville/

Finlay, L. (1994, August 19). Barley's Brew Pub serves up the suds from microbrewery. *Asheville Citizen Times*, p. 1C, 7C.

Flack, W. (1997). American microbreweries and neolocalism: "Ale-ing" for a sense of place. *Journal of Cultural Geography, 16*(2), 37–53.

Fraser, R. A., & Alonso, A. (2006). Do tourism and wine always fit together? A consideration of business motivations. In J. Carlsen & S. Charters (Eds.), *Global wine tourism: Research, management and marketing* (pp. 19–26). Cambridge, MA: CAB International.

Gallarza, M. G., Saura, I. G., & Garcia, H. C. (2002). Destination image: Towards a conceptual framework. *Annals of Tourism Research, 29*(1), 56–78.

Getz, D., & Brown, G. (2006). Critical success factors for wine tourism regions: A demand analysis. *Tourism Management, 27*(1), 146–158.

Glenn, A. F. (2010). Yay we win, now let's have a really big party. *Mountain Xpress.* Retrieved from http://mountainx.com/arts/art-news/060210yay_we_win_now_lets_have_a_really_big_party/

Glenn, A. F. (2012). *Asheville beer: An intoxicating history of mountain brewing.* Charleston, SC: American Palate.
Hall, C. M., Johnson, G., Cambourne, b., Macionis, N., Mitchell, R., & Sharples, L. (2009). Wine tourism: An introduction. In C. M. Hall, L. Sharples, B. Cambourne, & N. Macionis (Eds.), *Wine tourism around the world: Development, management and markets* (pp. 1–23). New York: Routledge.
Hatch, M. J., & Schultz, M. (2002). The dynamics of organizational identity. *Human Relations, 55*(8), 989–1018.
Hayward, S. D., & Jiang, D. S. (2016). Lunatics at the fringe: Teaching expository documentaries with Beer Wars. *The International Journal of Management Education, 14*(3), 388–410.
Hindy, S. (2014). *The craft beer revolution: How a band of microbrewers is transforming the world's favorite drink.* New York, NY: Palgrave Macmillan.
Howard-Grenville, J., Metzger, M. L., & Meyer, A. D. (2013). Rekindling the flame: Processes of identity resurrection. *Academy of Management Journal, 56*(1), 113–136.
Howley, M., & Westering, J. v. (2008). Developing wine tourism: A case study of the attitude of English wine producers to wine tourism. *Journal of Vacation Marketing, 14*(1), 87–95.
Jones, J. A. (2011). Who are we? Producing group identity through everyday practices of conflict and discourse. *Sociological Perspectives, 54*(2), 139–161.
Kiss, T. (1995, September 29). Highland to debut new brew. *Asheville Citizen-Times*, p. 1C, 9C.
Kiss, T. (2001, March 2). Welcome to the brew haha. *Asheville Citizen-Times*, p. 56.
Kiss, T. (2010, December 14). Brewinginashville1 [video file]. Retrieved from https://www.youtube.com/watch?v=aj_Mduhf0XE
Kohler, S. (2014). *2014 Asheville visitor profile: An inside look at the overnight leisure travel market in Asheville, NC.* Retrieved from http://www.ashevillecvb.com/wp-content/uploads/2014-Asheville-Visitor-Profile-09212015.pdf
Krug, E. C. (2010). *Asheville: Beer city USA.* Retrieved from http://intelligent-travel.nationalgeographic.com/2010/06/04/asheville_beer_city_usa_1/
Lamertz, K., Foster, W. M., Coraiola, D. M., & Kroezen, J. (2016). New identities from remnants of the past: An examination of the history of beer brewing in Ontario and the recent emergence of craft breweries. *Business History, 58*(5), 786–828.
Lamertz, K., Heugens, P., & Calmet, L. (2005). The configuration of organizational images among firms in the Canadian beer brewing industry. *Journal of Management Studies, 42*(4), 817–843.
Lyons, M. (n.d.). *Asheville brews cruise – Brewery tour.* Retrieved from http://www.ashevillenow.com/restaurants/asheville-beer/asheville-brews-cruise/
MacCannell, D. (1973). Staged authenticity: Arrangements of social space in tourist settings. *American Journal of Sociology, 79*(3), 589–603.

Martin, A., & Haugh, H. M. (1999). The malt whisky trail: The tourism and marketing potential of the whisky distillery visitor centre. *International Journal of Wine Marketing, 11*(2), 42–52.

McBoyle, G., & McBoyle, E. (2008). Distillery marketing and the visitor experience: A case study of Scottish malt whisky distilleries. *International Journal of Tourism Research, 10*(1), 71–80.

McGahan, A. M. (1991). The emergence of the national brewing oligopoly: Competition in the American market, 1933–1958. *The Business History Review, 65*(2), 299–284.

McGee, M. (1998). From head to foot. *Mountain Xpress*. Retrieved from http://mountainx.com/arts/art-news/0916brewgrass-php/

Myers, E. L. (2012). *North Carolina craft beer & breweries*. Winston-Salem, NC: John F. Blair, Publisher.

Penland, K. (2016). *5th annual AVL beer week launches May 27–June 4, 2016*. Retrieved from http://avlbeerweek.com/wp-content/uploads/sites/7/2016/05/FOR-IMMEDIATE-RELEASE-AVL-BEER-WEEK-2016.pdf

Prentice, R., & Andersen, V. (2003). Festivals as creative destination. *Annals of Tourism Research, 30*(1), 7–30.

Ram, Y., Bjork, P., & Weidenfeld, A. (2016). Authenticity and place attachment of major visitor attractions. *Tourism Management, 53*, 110–122.

Romanelli, E., & Khessina, O. M. (2005). Regional industrial identity: Cluster configurations and economic development. *Organization Science, 16*(4), 344–358.

Scherer, M. L. (2007). Building a bohemian boomtown: The construction of a 'creative class' in Asheville, North Carolina. *Sociation Today, 5*(2). http://www.ncsociology.org/sociationtoday/v52/asheville.htm

Schnell, S. M., & Reese, J. F. (2003). Microbreweries as tools of local identity. *Journal of Cultural Geography, 21*(1), 45–69.

Starnes, R. D. (2005). *Creating the land of the sky: Tourism and society in Western North Carolina*. Tuscaloosa, AL: The University of Alabama Press.

Stokes, R. (2008). Tourism strategy making: Insights to the events tourism domain. *Tourism Management, 29*(2), 252–262.

Strom, E., & Kerstein, R. (2015a). Mountains and muses: Tourism development in Asheville, North Carolina. *Annals of Tourism Research, 52*, 134–147.

Strom, E., & Kerstein, R. (2015b). The homegrown downtown: Redevelopment in Asheville, North Carolina. *Urban Affairs Review*, 1–27.

Wang, S., & Chen, J. S. (2015). The influence of place identity on perceived tourism impacts. *Annals of Tourism Research, 52*, 16–28.

CHAPTER 12

An Exploration of the Motivations Driving New Business Start-up in the United States Craft Brewing Industry

Erol Sozen and Martin O'Neill

Introduction

In the period from 1980 to 1999, 34 million new jobs were created in the United States, and it is suggested that entrepreneurs were responsible for creating 95 % of this new wealth (Timmons & Spinelli, 1999). It is not surprising then that academics, potential entrepreneurs, and policy makers alike have an interest in the factors driving this form of economic endeavor. Nica, Grayson, and Gray (2015, p. 119) posit that a "strong relationship has been found to exist between entrepreneurial activity and economic growth", and that "all major theories of entrepreneurship view entrepreneurs as creators of new economic activity, which leads to wealth creation". Consequently, entrepreneurial engagement remains a common policy thread of most government bodies charged with economic development and the strengthening of inter-sectoral economic linkages (Mazzarol, Volery, Doss, & Thein, 1999). Viewed as a form of economic

---

E. Sozen (✉) • M. O'Neill
Nutrition, Dietetics, & Hospitality Management,
Auburn, AL, USA

panacea, entrepreneurial activity is widely encouraged as a means of breathing new life and hope into lesser and underdeveloped communities and regions, creating jobs, generating income, much-needed innovative spirit, and knowledge capital (Amorós, 2009; Dyck & Ovaska, 2011; Stephan, Hart, & Drews, 2015).

One such form of economic endeavor is that of craft brewing, which has seen exponential growth in recent years in both urban and rural areas. Craft brewers are not just manufacturers providing employment opportunity, but many serve as destinations in their own right, generating multiple incomes effects, increasing the local tax base and strengthening inter-sectoral linkages as well as supply chain opportunities (International Economic Development Council, 2016). According to the Brewers Association, in 2014, the industry contributed over $55 billion to the US economy, and an estimated 424,000 were employed directly or indirectly in the industry (Brewers Association, 2015).

The Association defines craft brewers as small, independent, and traditional. Thus, craft brewing combines rigorous science (fermentation, microbiology, and in some instances, cold filtering techniques) with wide variation based on personal taste, brewing method, marketplace demand, and indigenous or imported ingredients. Between 2008 and 2011, of the 720 new craft breweries opened in the United States, it is estimated that 22.2 % closed within the first five years of business (Watson, 2014). This combination of elements makes potential engagement in this industry especially challenging to the entrepreneurially minded amateur craft brewer (Murray & O'Neill, 2012). At a time when the craft brewing industry is on the rise and perceived in many locales as a form of economic salvation (Funari, 2013), it is appropriate that the motivations driving engagement in this sector be fully investigated so that the future of this creative form of economic endeavor can be sustained.

Results from the broader business sector point to a variety of motivations driving entrepreneurial engagement including, in a primary sense, a desire for independence and increased financial security, and, in a secondary sense, frustration with previous and current employers, redundancy, or business closure (Carsrud & Brännback, 2011; Nica et al., 2015). That being said, surprisingly little research has been undertaken to uncover the motivations driving new business start-up in the commercial craft brewing industry. While there is a paucity of research on the decision-making process that drives the consumer to this more expensive product line, as well as the challenges posed by a range of environmental factors upon business operation, little is known about the individual motivations driving entrepreneurial engagement. This chapter addresses the following research questions: first, what motivates business operators to engage entrepreneurially in the commercial world of craft brewing? And second, are these motivations any different than the motivations driving entrepreneurs in other business sectors?

Entrepreneurial Motivation

While the definition of entrepreneurship is still a matter of debate, three common denominators of entrepreneurial activity have come to dominate the literature, namely opportunities, uncertainty, and innovation (Nica et al., 2015). Debate has centered on the entrepreneur's ability to seek out, spot, create, and take advantage of opportunities. Some of this is situational, market uncertainty, and/or risk avoidance driven; in other words, reaction to forced circumstance or necessity driven (Kirzner, 2009; Rosa, Kodithuwakku, & Balunywa, 2006), for example, forced self-employment due to redundancy or health issues. However, in other circumstances, the entrepreneur utilizes their innovative and creative talents to identify or discover economic opportunity hitherto not available. They spot an opportunity in the market for a much-needed product or service (Schumpeter, 1966) and respond to it. According to Turkina and Thai (2015, p. 213), this opportunity-driven form of entrepreneurship is widely viewed as more desirable and "beneficial because, unlike necessity entrepreneurship, it is growth oriented and therefore enhances economic growth".

This raises the question: is this ability to spot or discover opportunity a skill that is endogenous to the entrepreneur or is it largely exogenous and a mere reaction to environmental circumstance? This may also help answer another question—why are some people drawn to entrepreneurial endeavors? While most studies on entrepreneurial motivation focus on factors such as prevailing economic circumstance, economic development strategies, and related policy initiatives that drive necessity entrepreneurship, they do little to explain opportunity entrepreneurship and the mindset of those engaged in creating or discovering new socio-economic opportunity (Thai & Turkina, 2012).

The authors suggest that other socio-psychological factors must be considered when it comes to opportunity-driven entrepreneurial motivation. They point to the importance of social psychology theories and methods in helping to understand what motivates opportunity-oriented entrepreneurs, positing that entrepreneurship is by definition a social phenomenon that begins with opportunity recognition and continues through an organizing process that involves a multitude of interactions with others, culminating in new venture creation developing around a culture created by its founder(s). The authors identify two models that tend to dominate the literature: those that study the entrepreneur such as Bandura's (1997) model of self-efficacy and Ajzen's (1996) Theory of Planned Behavior and those that account for broader societal factors and their influence upon entrepreneurial endeavor, such as Thurik and Dejardin's (2012) model of social legitimization. Central to both are the concepts of desirability, feasibility, and the social context within which entrepreneurial activity might occur. While would-be entrepreneurs might have the desire to engage in entrepreneurial endeavors, actual behavior is driven by the feasibility of operation which is largely shaped by factors beyond their control in the broader social environment. This raises the question of control and the associated risk attached to new venture creation when there is a perceived lack of control, thereby impeding potential success of the venture. What is clear is that the concept of entrepreneurial motivation cannot be studied as a static concept, rather it needs be looked at as a process that is subject to both individual (endogenous) and broader societal (exogenous) issues. It is only by accounting for all of the factors that might accelerate or impede entrepreneurial endeavor that policy makers and entrepreneurial prospects might better be prepared for long-term success.

This view is shared by Bygrave (1989, p. 21) who states that entrepreneurship is a "process of becoming rather than a state of being", which speaks to the ongoing and continuous nature of the task, from opportunity recognition through to actual operation and long-term, sustainable growth. This also implies a degree of environmental awareness, scanning, interaction with multiple constituency groups and an ability to respond effectively and in a timely manner to issues that might influence performance and long-term growth. So, what are these external factors that might challenge new venture creation and traits that define the more successful entrepreneur?

Motivational Influence

Research on the potential influence of the broader environment has sought to investigate the interplay between the prospective entrepreneur and those external factors that influence conceptual development, process development, business planning, process decision-making, and longer-term success, in other words, their influence upon the start-up decision and operational decisions thereafter. These factors relate to the political, economic, socio-cultural, and technological (PEST) environments and their ability to accelerate or impede new business start-up (Aguilar, 1967). Shapiro and Stiglitz (1984) cite the influence of societal attitudes as well as the prevailing economic climate. Further, Learned (1992) suggests that the environment impacts the opportunity culture that may stimulate the entrepreneurs' intentions, while others (Carsrud, Gaglio, & Olm, 1986) highlight the influence of supportive business networks and the prevailing economic development approach of local government. The craft brewing industry is unique in that religious, moral, and legal issues are added to the operational challenges that face any new business/entrepreneurial venture (Baginski & Bell, 2011).

At the individual level, we are dealing with the founder of the business, the person who has the "light bulb" moment and navigates the process to opening the doors on actual operations. This is the visionary who translates the idea into a company vision, mission, and value statements, creates an organizational culture, and engages in hiring and leading like-minded individuals in pursuit of this vision and mission. The founder is also responsible for navigating the wider socio-political and economic environment to secure the capital and know-how necessary to breathe life into the project, while minimizing or managing the associated risks sufficiently to go into business. Mazzarol et al. (1999, p. 48) put it this way, "before organizations there are pre-organizations that exist initially only as the thoughts, ideas, or dreams of the would-be entrepreneur. Through the startup process, the founder's thoughts are sometimes translated in a preorganization (an attempt to found), and then sometimes (but not always), an organization". A number of studies have addressed a multitude of personal influences, for example, demographic factors such as age and gender (Stephan et al., 2015); background factors such as education, employment history, and family and financial background (Uygun & Kasimoglu, 2013); and motivational factors or personal rewards such as the need for personal control, independence, and freedom (Birley & Westhead, 1994; Fayolle, Linan, & Moriano, 2014; Shane, Locke, & Collins, 2003), the need for achievement and creative endeavor (Fillis & Rentschler, 2010; Ward, 2004), and opportunity orientation (Short, Ketchen, Shook, & Ireland, 2010). Carsrud and Brännback (2011, p. 11) tell us that the study of motivation is important as it provides answers to a number of questions, including: "what activates a person, what makes an individual chose one behavior over another, and why do different people respond differently to the same motivational stimuli?" The authors also highlight three important aspects of motivation: "activation, selection-direction and preparedness of response". In turn, these serve to differentiate between what have been termed "drive" or "push" theories and "incentive" or "pull" theories of entrepreneurial motivation. A recent audit completed by the UK-based Enterprise Research Center (2015) serves to further understand the key drivers of entrepreneurial motivation and what consequences different entrepreneurial motivations have had on entrepreneurial performance.

While the evidence presented in response to all three questions is relevant to the subject of inquiry, the focus on the multi-dimensional typologies of entrepreneurial motivation is deemed most relevant to this chapter. The report reviewed a total of 27 research studies making use of reliable multi-item indices and factor analyses as a statistical technique to derive the underlying dimensionality of the entrepreneurial motivation construct. Findings ranged from a two-factor solution (Gorgievski, Ascalon, & Stephan, 2011) to a seven-factor solution (Rouse & Jayawarna, 2011) indicating a wide degree of variance in explaining the entrepreneurial motivation construct and the weighted importance of certain dimensions over others. That being said, the report identified a range of motivational factors that were found to be common across most studies reviewed including the need for achievement, challenge and learning, independence and autonomy, income security and financial success, personal and societal recognition and success, continuity of family tradition, dissatisfaction with current or prior work experience, and community and social motivations.

Two studies in particular stood out from the report. First, the work of Scheinberg and Macmillan (1988) who received responses from 1402 founder entrepreneurs revealing a range of different motivations and significant differences in these motivations across 11 different countries including the United States, Australia, China, Puerto Rico, and central and northern European countries. Scheinberg and Macmillan (1988) presented respondents with a list of 38 different potential entrepreneurial motivators and had them scale the extent to which each motivation influenced their decision to become an entrepreneur. The data was then subjected to a principal components factor analysis which ultimately presented good to strong reliabilities across six different dimensions. These they labeled: Need for Approval, Perceived Instrumentality of Wealth, Communitarianism, Need for Personal Development, Need for

Independence, and Need for escape. The data was subsequently exposed to a cluster analysis which turned up very different weightings for each factor by country which very much speaks to the influence of the wider social and business context upon new venture creation in each country. The second study worthy of consideration is that by Birley and Westhead (1994) which addressed the issue of motivation for new business start-up in Great Britain. This study surveyed principal owner-managers of 405 new independent businesses and sought to glean answers to two central research questions: are there any differences in the reasons that owner-managers articulate for starting their businesses? And if there are, do they appear to affect the subsequent growth and size of the business? While results were found to be both reliable and valid and seemed to confirm the earlier results of Scheinberg and Macmillan's (1988) study, two further components were identified. These were labeled "Tax Reduction and Indirect Benefits" and "Follow Role Models" as identified by Dubini (1989). Both factors were found to be reliable explaining a unique proportion of the variance in the data set. For this reason, their instrument was to serve as a foundation for the present study in that it helped inform thinking on the long list of potential motivators for entry into the competitive world of craft brewing.

METHODS

This study explores the motivations driving entrepreneurial engagement in the US craft brewing industry and employed a mixed methods approach. Qualitatively, personal interviews were conducted with brewery owners at a variety of locations in the southeastern United States. This region and these breweries were chosen based upon proximity to the authors' place of employment. Cost and time associated with travel were considerations here. While a total of 15 owners were contacted in relation to their willingness to participate in the study, only five gave their consent. At the point of initial contact, all five had been in business for at least three years. All interviews were conducted in person, with responses recorded for later transcription and analysis. Each interview lasted approximately 30–45 minutes and interviewees were asked about a number of themes including past association with brewing and beer, navigation of the new business start-up process, motivations for engagement, and the daily challenges regarding business operation and growth. These questions and themes were found common to previous studies conducted on entrepreneurial motivation in the broader business sector (Birley & Westhead, 1994; Carsrud & Brännback, 2011; Nica et al., 2015; Scheinberg & Macmillan, 1988).

These responses as well as the scale developed and tested by Birley and Westhead (1994) served as a foundation for the development of the more quantitative element of the research process; namely, the development of an online survey that was distributed through the American Brewers Association to 2456 craft breweries across the United States. The survey comprised a number of distinct sections and addressed a variety of themes including respondent demographics, entrepreneurial motivation, entrepreneurial orientation, work/life satisfaction, and business challenges. That said, the ensuing analysis addresses respondent demographics and entrepreneurial motivation only as the key themes for this chapter. As indicated, the survey relied upon the earlier work of Birley and Westhead (1994) to assess entrepreneurial motivation. Their study was found to have strong reliability and validity indicators and perfectly suited to the current investigation. Their study identified seven different entrepreneurial motivators including (1) the need for approval, (2) need for independence, (3) need for personal development, (4) welfare considerations, (5) perceived instrumentality of wealth, (6) tax reduction and indirect benefits, and (7) the need to follow role models. This was adapted and contextualized to the craft brewing audience with the inclusion of one additional variable, which was added based upon feedback provided by expert informants during the preliminary phase of the study. Variables were measured on a five-point Likert-type agreement scale anchored at (1) strongly disagree to (5) strongly agree.

To minimize the potential for ambiguity of the survey questions, a pilot study was undertaken, wherein the instrument was administered to members of a local home-brewing club and five founding craft brewers. Feedback was solicited on the relevance and clarity of questions, scaling techniques used, construct validity, and the time required to complete the survey. The feedback led to a number of refinements, grammatical modifications, and the addition of one more variable on the entrepreneurial motivation scale (EMS), which was "to make beers that aligned with my tastes". This variable was thought to be relevant to the subject of inquiry. The finalized questionnaire was then shared with the Brewers Association for further input. Upon review, a number of additional changes (mostly grammatical) were requested and the questionnaire was distributed to a convenience sample of US craft brewers through the Brewers Association Brew Forum Blog. This blog is shared with commercially oriented craft brewers nationally. Potential respondents were provided with a little background on the nature and intent of the survey and then invited to click on a link to take the survey. They were then brought to an informed consent letter and provided with the option to proceed or not proceed with the completion of the survey. For the purpose of this study, sample respondents were defined as founders or business partners in the craft brewing venture. All responses were collected online using Qualtrics Software. The survey was left open over a six-week period spanning March to mid-April 2013, with reminder emails sent through the Brewers Association at weeks three and five.

Sample Characteristics

A total of 213 valid responses were received over the six-week period the survey remained open representing a response rate of approximately 9 % of all registered breweries with the Brewers Association. Upon further analysis, however, the authors were concerned that not all of the respondents registered as owner operators with 95 respondents (44 %) identifying as employees of their respective breweries. On reflection, the inclusion of a selection/deselection question would have helped avoid this issue. This left a total of 118 respondents who identified as either owner operators or business partners. For the purposes of further analysis, data was recoded so that only those responses received from owner operators or business partners would be analyzed. Table 12.1 shows that of the 118 respondents who self-identified as owner operators or partners, 89 % were male, 92 % were Caucasian and equally dispersed throughout the United States. Some 72 % of respondents were over the age of 34 with "Generation X" respondents accounting for 49 % of the response rate. Almost 63 % of responding breweries were located in either "inner city" (31 %) or "suburban" areas (31 %), with 27 % of respondent breweries being located in "small towns" and 9 % in "rural areas". A variety of reasons were cited for choice of business location including "ease of distribution" (20 %), "hometown" (19 %), and "cost" (18 %). The dominant (21%) income level was recorded at $76–$99,000 per annum, with approximately 92 % of respondents earning in excess of $40,000 in the year of the study. In terms of educational background, a majority of respondents (just over 54 %) declared that they had earned a bachelor's degree and 34 % of respondents declared that they held a masters (23 %) or doctoral degree (11 %). Just under 58 % of the respondents indicated that they had been in busi-

Table 12.1 Demographic background

Demographics	N	%	Demographics	N	%
Generations			**Brewing qualification**		
19–33 Millennium Gen.	33	28.0	Yes	32	27.6
34–49 Gen X	58	49.2	No	184	72.4
50–67 Baby boomer	25	21.2	**Years in business**		
68 and older	2	1.7	Less than 1 year	29	27.1
Gender			1–3 years	33	30.8
Female	13	11.0	4–5 years	14	13.1
Male	105	89.0	6–10 years	7	6.5
Ethnicity			More than 11 years	24	22.4
Pacific Islander	1	0.8	**Years being home brewer**		
Caucasian	109	92.4	Less than 1 year	2	21.0
Asian	2	1.7	1–5 years	1	2.0
Hispanic	2	1.7	6–10 years	7	7.8
Multi-racial	2	1.7	11–15 years	9	9.3
I prefer not to answer	2	1.7	More than 15 years	28	8.9
Income			**Operating tasting facility**		
Under $25,000	5	4.3	Yes	84	71.0
$25,000–$39,000	4	3.4	No	21	29.0
$40,000–$54,000	11	9.5	**Fee tour/tasting session**		
$55,000–$75,000	20	17.2	Yes	34	28.0
$76,000–$99,000	28	24.1	No	66	72.0
$100,000–$150,000	22	19.0			
Over $150,000	26	22.4			
Education					
Training/apprenticeship	1	0.8			
Some college	7	5.9			
Associate's degree	6	5.1			
Bachelor's degree	64	54.2			
Master's degree	27	22.9			
Doctoral degree	13	11.0			

ness for under 3 years, while 42 % indicated that they had been in business for 3 years or more. Of this latter group, almost 23 % had been in business for more than 11 years. Fewer than 28 % of respondents had any formal educational background in brewing. Of those that did, they were mostly (40 %) earned through private institutes (Siebel Institute, the Institute of Brewing and Distilling) or the American Brewers Guild. Some 25 % of those that responded indicated that they had earned a formal university qualification. Approximately 80 % of respondents indicated that this was a second career choice with most declaring that they had previously worked in an engineering-, science-, business-, or hospitality-related profession. Some 83 % of respondents declared a history with home brewing prior to engaging with craft brewing in a commercial sense. Some 71 % of respondents indicated that they operated a tasting facility, with 56 % of those that did indicating that they offered free tastings and 72 % declaring that they also offered free educational tours.

Descriptive Analysis

Table 12.2 summarizes the mean, standard deviation, and skewness for each of the EMS variables. The variables are grouped around the respective factor (dimensions) uncovered in the original Birley and Westhead (1994) study. The results point to, on average, a high level of agreement for most variables with a range spanning a low $m = 1.96$ for variable 22 (as a vehicle to reduce the burden of the taxes I face) to a high $m = 4.38$ for variable 7 (to have considerable freedom to adapt may own approach to my work) on the five-point agreement scale. When looked at from a dimension perspective, those factors pertaining to the "Need for Independence", "Personal Development", and "Perceived Instrumentality of Wealth" appear as strong motivators for entrepreneurial engagement.

Performance of the Research Instrument

While the overriding goal of this research was to identify the key motivators driving entrepreneurial engagement, it was also deemed essential to test the psychometric properties of the research instrument for reliability. Reliability analyses were conducted on the scale as it related to all respondents ($n = 213$) and for those declaring as entrepreneurs only ($n = 118$). The scale performed well with the entire respondent group ($\alpha = 0.88$, $n = 213$) and the entrepreneurial subset ($\alpha = 0.86$, $n = 118$). These reliability scores clearly exceed the usual recommendation of $\alpha = 0.70$ for establishing internal consistency of the scale.

The EMS was then exposed to an exploratory factor analysis using the principal component extraction technique. This was designed to attest to the scales' ability to discriminate between the variables explaining the underlying factor structure and, by definition, the key motivators driving entrepreneurial engagement in craft brewing among this respondent group. The analysis used the VARIMAX factor rotation procedure

Table 12.2 Entrepreneurial motivation variables

Dimensions/variables	Mean	Std. dev.	Skewness
Need for approval			
v7—Desire to have high earnings	2.43	1.109	0.442
v9—To achieve something and to get recognition for it	3.89	0.906	−0.589
v12—To achieve a higher position for myself in society	2.56	1.110	0.312
v16—To increase the status and prestige of my family	2.15	1.015	0.624
v18—To be respected by friends	3.02	0.986	−0.323
v21—To have more influence in my community	3.07	1.119	−0.256
Need for independence			
v1—To have considerable freedom to adapt my own approach to work	4.38	0.779	−1.470
v3—To control my own time	3.78	1.066	−0.747
v4—It made sense at that time in my life	4.13	0.931	−1.149
v6—To have greater flexibility for my personal and family life	3.17	1.240	−0.097
Need for personal development			
v10—To continue learning	4.22	0.704	−0.497
v13—To be innovative and to be in the forefront of technological development	3.41	1.059	−0.433
v14—To develop an idea for a product	3.00	0.907	−1.213
v2—To take advantage of an opportunity that appeared	4.28	0.854	−1.346
v24—To make beers that align with my tastes	*4.10*	*0.916*	*−0.906*
Welfare consideration			
v11—To contribute to the welfare of my relatives	2.65	1.195	0.271
v17—To contribute to the welfare of the community that I live in	3.88	0.880	−0.701
v19—To contribute to the welfare of people with the same background as me	2.77	1.031	0.078
Perceived instrumentality of wealth			
v8—To be challenged by the problems and opportunities of starting and growing a new business	4.06	0.834	−0.676
v5—To give myself, my spouse, and children security	2.89	1.251	0.119
Tax reduction and indirect benefits			
v15—To have access to indirect benefits such as tax exemptions	2.23	1.147	0.444
v20—As a vehicle to reduce the burden of taxes I face	1.96	1.013	0.800
Follow role models			
v22—To follow the example of the person that I admire	2.58	1.151	0.140
v23—To continue a family tradition	1.97	1.122	0.980

in SPSS 22. A component matrix was initially generated to ensure that the analyzed variables had reasonable correlations (greater than or equal to 0.5) with other variables. Unrotated and rotated component matrices were inspected and all variables were found to correlate well. The result of the corresponding KMO of "sampling adequacy" was 0.710 and Bartlett's test for sphericity was 1001.354, significant at the level of 1 %(sig. = 0.001). The results of these tests rendered the data factorable and consequently the factor analysis was generated.

Table 12.3 highlights clean factor loadings across eight different dimensions, explaining 69 % of the variance. Factor 1 has been labeled "TAX"

Table 12.3 Exploratory factor analysis—EMS

Variables	F1 TAX	F2 FAMILY	F3 PD	F4 COMM	F5 PI	F6 OPP	F7 APP	F8 ?
v.20	0.849							
v.15	0.765							
v.23	0.674							
v.16	0.614							
v.5		0.741						
v.7		0.726						
v.6		0.670						
v.11		0.606						
v.13			0.763					
v.9			0.644					
v.14			0.631					
v.12			0.550					
v.10			0.514					
v.21				0.807				
v.17				0.703				
v.19				0.554				
v.3					0.815			
v.1					0.705			
v.2						0.736		
v.4						0.711		
v.18							0.669	
v.8							−0.639	
v.24								0.713
v.22								
Eigenvalue	6.183	2.339	1.809	1.463	1.408	1.219	1.100	1.021
% of variation	25.763	9.744	7.537	6.094	5.867	5.078	4.582	4.252
α	0.81	0.76	0.72	0.69	0.62	0.40	−0.32	–

and seems somewhat reflective of Birley and Westhead's "Tax reduction and indirect benefits" dimension (see Table 12.2 for this and subsequent comparisons). Factor 2, "FAMILY", appears to be an amalgamation of the "Need for approval" and "Independence" dimensions. Factor 3, "PD" for personal development, aligns very well with their "Need for Personal Development" dimension, and Factor 4, "COMM" for community, also aligns very closely with the dimension labeled "Welfare Consideration". Factor 5, "PI" for personal independence, also aligns with the "Need for Independence" dimension. Factor 6, "OPP" for opportunity, seems to be an amalgamation of the original "Need for Personal Development" and "Need for Independence" dimensions and Factor 7, "APP" for approval, loosely corresponds to the "Need for Approval" dimension from the original study. Factor 8, labeled "?" pertains solely to the additional variable the researchers sought to include in the study believing it would factor into the "Need for Independence" dimension. Clearly, this did not factor out as imagined. Table 12.3 also highlights strong-to-moderate reliability co-efficients (α) for factors 1 through 5 ranging from $\alpha = 0.81$ for factor 1 to $\alpha = 0.69$ for factor 5, and low to weak reliabilities for factors 6 and 7. It was not possible to calculate a reliability co-efficient for factor 8.

Discussion, Limitations, and Conclusions

The aim of this study was to investigate the motivations that drive entrepreneurial engagement in the US craft brewing industry and to ascertain if these motivations are similar to those of entrepreneurs in other business sectors. The work has both practical and theoretical implications for policy makers, craft brewers, and academics. Not least for those seriously considering entry into what some now consider a very saturated marketplace (Morris, 2015; Scully, 2016). From a policy perspective, while much has been written on the psychological underpinnings of entrepreneurial motivation in the broad business sector, little if any work has been completed on this topic in the craft brewing industry. Given recent growth in this industry nationally and its ability to generate jobs, income, and to revitalize otherwise depressed inner city, urban, and rural environments, policy makers at the local, regional, and national levels are interested in creating a favorable environment for craft breweries to set up and do business. An understanding of those factors driving the start-up process, their relative importance to entrepreneurs and an ability to target resources, financial and other, to ease this process is critical.

From a research perspective, the results align favorably with other studies that have been conducted in the broad business field and support the contention that entrepreneurs in the craft brewing industry appear to be driven by similar start-up motivations to those of entrepreneurs in other business sectors. The results identify five components that correspond almost exactly to the work of Birley and Westhead (1994). That said, the relative weightings are somewhat different, and there is a degree of overlap with certain factors. Interestingly, the issue of "tax reduction and indirect benefits" (TAX) as well as family (FAM) and financial security appear most important to respondents, whereas the "need for approval" (PA) and "independence" (PI) appeared most important in Birley and Westhead's study. The issue of tax reduction is important as it is also supported by Dubini's (1989) earlier work, which spoke to the desire to increase personal wealth by retaining as much earned money as possible. Certain of these factors were also identified in earlier studies such as the "need for independence" which corresponds to Scheinberg and Macmillan's (1988) 11-country study of entrepreneurial motivation. Factor 7, titled "APP" for approval corresponds with McClelland's (1956) earlier work on the need for achievement; however, it was found to be not so critical to this group. As with other studies, this does not imply mutual exclusivity, rather it points to the myriad of motivational influences that may be at play at any one point in time and their ability to help policy makers and potential investors determine which ventures are likely to be successful over the longer term.

The results further speak to the importance of what Shane et al. (2003, p. 279) term "Human Agency", where entrepreneurial activity "depends on the decisions that people make, suggesting that the attributes of the decision makers should influence the entrepreneurial process". As such, the further development of entrepreneurial theory requires ample consideration of the factors driving people as founders to make entrepreneurial decisions. The study also attests to the reliability of Birley and Westhead's (1994) instrument for measuring entrepreneurial motivation. The instrument demonstrated good reliability. Moving forward it would be useful to test for variance between different groups of entrepreneurs in the brewing field based upon gender, entrepreneurial orientation, founders versus investors, new entrepreneurs versus well-established entrepreneurs and perhaps primary versus secondary career choice.

Limitations of the study include the inability to exclude employees as opposed to the supposed target group, namely business owners/operators only from the total sample. It would also have proven beneficial to ask as to the primary motivation for engaging in craft brewing; in other words, whether the decision was driven by desire or circumstance. The ability to analyze responses based upon whether respondents were opportunity seekers versus circumstantial entrepreneurs may have proved interesting.

REFERENCES

Aguilar, F. J. (1967). *Scanning the business environment*. London: Macmillan.

Ajzen, I. (1996). The directive influence of attitudes on behavior. Retrieved December 12, 2016, from psycnet.apa.org

Amorós, J. E. (2009). Entrepreneurship and quality of institutions: A developing-country approach (No. 2009.07). Research paper/UNU-WIDER.

Baginski, J., & Bell, T. L. (2011). Under-tapped?: An analysis of craft brewing in the Southern United States. *Southeastern Geographer, 51*(1), 165–185.

Bandura, A. (1997). *Self-efficacy: The exercise of control*. New York: Freeman.

Birley, S., & Westhead, P. (1994). A taxonomy of business start-up reasons and their impact on firm growth and size. *Journal of Business Venturing, 9*(1), 7–31.

Brewers Association. (2015). *Craft beer sales by state*. Retrieved October 22, 2016, from https://www.brewersassociation.org/statistics/by-state/?state=PA

Bygrave, W. D. (1989). The entrepreneurship paradigm (I): A philosophical look at its research methodologies. *Entrepreneurship Theory and Practice, 14*(1), 7–26.

Carsrud, A., & Brännback, M. (2011). Entrepreneurial motivations: What do we still need to know? *Journal of Small Business Management, 49*(1), 9–26.

Carsrud, A., Gaglio, C., & Olm, K. (1986). Entrepreneurs, mentors, networks, and successful new venture development. *American Journal of Small Business, 12*(2), 13–18.

Dubini, P. (1989). The influence of motivations and environment on business start-ups: Some hints for public policies. *Journal of Business Venturing, 4*(1), 11–26.

Dyck, A., & Ovaska, T. (2011). Business environment and new firm creation: An international comparison. *Journal of Small Business & Entrepreneurship, 24*(3), 301–317.

Enterprise Research Center. (2015). Understanding motivations for entrepreneurship. Retrieved October 10, 2016, from https://www.gov.uk/government/uploads/system/uploads/attachment_data/file/408432/bis-15-132-understanding-motivations-for-entrepreneurship.pdf

Fayolle, A., Linan, F., & Moriano, J. A. (2014). Beyond entrepreneurial intentions: Values and motivations in entrepreneurship. *International Entrepreneurship and Management Journal, 10*(4), 679–689.

Fillis, I., & Rentschler, R. (2010). The role of creativity in entrepreneurship. *Journal of Enterprising Culture, 18*(1), 49–81.

Funari, C. (2013). *Craft beer's impact on the local economy discussed*. Brewbound: BEVNET.

Gorgievski, M. J., Ascalon, M. E., & Stephan, U. (2011). Small business owners' success criteria, a values approach to personal differences. *Journal of Small Business Management, 49*(2), 207–232.

International Economic Development Council. (2016). The craft beverage boom – Capture the economic opportunity, IEDC Webinars.

Kirzner, I. M. (2009). The alert and creative entrepreneur: A clarification. *Small Business Economics, 32*(2), 145–152.

Learned, K. E. (1992). What happened before the organization? A model of organization formation. *Entrepreneurship: Theory and Practice, 17*(1), 39–49.

Mazzarol, T., Volery, T., Doss, N., & Thein, V. (1999). Factors influencing small business start-ups: A comparison with previous research. *International Journal of Entrepreneurial Behaviour & Research, 5*(2), 48–63.

McClelland, D. C. (1956). Risk-taking in children with high and low need for achievement (RPRT). DTIC Document.

Morris, C. (2015). Is craft beer in a bubble? *Fortune*. Retrieved February 3, 2017, from http://fortune.com/2015/05/16/is-craft-beer-in-a-bubble/

Murray, D. W., & O'Neill, M. a. (2012). Craft beer: Penetrating a niche market. *British Food Journal, 114*(7), 899–909.

Nica, M., Grayson, M., & Gray, G. T. (2015). Taxonomy of the determinants of entrepreneurial activity. *Journal of Economics and Economic Education Research, 16*(3), 119.

Rosa, P. J., Kodithuwakku, S., & Balunywa, W. (2006). Entrepreneurial motivation in developing countries: What does "necessity" and "opportunity" entrepreneurship really mean? *Frontiers of Entrepreneurship Research, 26*(20), 1–14.

Rouse, J., & Jayawarna, D. (2011). Structures of exclusion from enterprise finance. *Environment and Planning C: Government and Policy, 29*(4), 659–676.

Scheinberg, S., & Macmillan, I. (1988). *An eleven-country study of the motivation to start a business: Frontiers of entrepreneurship research*. Wellesley, MA: Babson College.

Schumpeter, J. (1966). *Invention and economic growth*. Cambridge, MA: Harvard University Press.

Scully, T. (2016). Craft brewing, dilution, saturation, or opportunity? *Wine and Craft Beverage News*. Retrieved February 3, 2017, from http://wineandcraftbeveragenews.com/craft-brewing-dilution-saturation-or-opportunity/

Shane, S., Locke, E. A., & Collins, C. J. (2003). Entrepreneurial motivation. *Human Resource Management Review, 13*(2), 257–279.

Shapiro, C., & Stiglitz, J. E. (1984). Equilibrium unemployment as a worker discipline device. *The American Economic Review, 74*(3), 433–444.

Short, J. C., Ketchen, D. J., Shook, C. L., & Ireland, D. (2010). The concept of "opportunity" in entrepreneurship research: Past accomplishments and future challenges. *Journal of Management, 36*(1), 40–65.

Stephan, U., Hart, M., & Drews, C.C. (2015). *Understanding motivations for entrepreneurship: A review of recent research evidence*. Enterprise Research Center.

Thai, M. T. T., & Turkina, E. (2012). *Entrepreneurship in the informal economy: Models, approaches and prospects for economic development*. London: Routledge.

Thurik, R., & Dejardin, M. (2012). Chapter 14: Entrepreneurship and culture. In *Entrepreneurship in Context* (Vol. 3, p. 175). London: Routledge.

Timmons, J., & Spinelli, S. (1999). New venture creation: Entrepreneurship for the 21st century. Retrieved November 12, 2016, from http://s3.amazonaws.com/academia.edu.documents/31162458/MBA-559-4.pdf?AWSAccessKeyId=AKIAJ56TQJRTWSMTNPEA&Expires=1483033555&Signature=NwfBQL%2FwSSsg3uELWPLeIspMm7Y%3D&response-content-disposition=inline%3B%20filename%3DNew_venture_creation_Entrepreneurship_fo.pdf

Turkina, E., & Thai, M. T. T. (2015). Socio-psychological determinants of opportunity entrepreneurship. *International Entrepreneurship and Management Journal, 11*(1), 213–238.

Uygun, R., & Kasimoglu, M. (2013). The emergence of entrepreneurial intentions in indigenous entrepreneurs: The role of personal background on the antecedents of intentions. *International Journal of Business and Management, 8*(5), 24–40.

Ward, T. B. (2004). Cognition, creativity and entrepreneurship. *Journal of Business Venturing, 19*(2), 173–188.

Watson, B. (2014). Closings signal competition not failure. *Brewers Association*. Retrieved November 12, 2016, from https://www.brewersassociation.org/insights/closings-signal-competition-not-problems/

CHAPTER 13

Conclusion

Susan L. Slocum, Christina T. Cavaliere, and Carol Kline

The understanding and study of sustainability is ever emerging and fittingly dynamic. The subtlety of sustainability within the craft beverage industry is further illustrated in this volume through theoretical and applied research results. Geographic case studies elucidate the nuances of sustainability brought forth in the craft beveragescape. Additionally, aspects of sustainability specific to the craft beverage sector are brought to light within this collection. In Chapter 1, Slocum situates the definition of sustainability for the purposes of this volume as "the balanced relationship of behavioral

S.L. Slocum (✉)
Tourism and Event Management, George Mason University,
Manassas, VA, USA

C.T. Cavaliere
Hospitality and Tourism Management Studies,
Galloway, NJ, USA

C. Kline
Appalachian State University, Walker College of Business,
Boone, NC, USA

© The Author(s) 2018
S.L. Slocum et al. (eds.), *Craft Beverages and Tourism, Volume 2*,
DOI 10.1007/978-3-319-57189-8_13

Common Themes

Quality Sourcing and Ingredients

Ingredients emerge as an important-to-consumer trend in the craft beverage sector. For example, in Chapter 8, Curtis, Bosworth, and Slocum found that drink tourists appreciate local food experiences and have preferences for locally sourced food and beverages via their interests in farmers' markets and community-supported agricultures. Moreover, in Chapter 3, Graefe, Mowen, and Graefe identify the value of locally sourced ingredients in Pennsylvania breweries, showing that consumers supported both local sourcing and the craft breweries' engagement with environmental activism. Myles and Breen, in Chapter 10, also discuss the role that fresh ingredients play within beer production, which are valued as a key component to quality beer production. They acknowledge that local sourcing reinforces the message of quality and freshness.

Ethical and sustainable practices can go far beyond the ingredients of the beverages themselves to create resonating green supply chain management. Jones shows, in Chapter 2, that the craft brewers he investigated used some of the most ethically reputable companies within their supply chain as a means to ensure quality results in their manufacturing process. However, there seems to be a lack of demand witnessed by the producers for organic beverage production. In fact, Jones found a line of organic beer that was actually discontinued due to underperforming sales. This lack of demand may change as niches within craft beverage continue to be developed.

Resource Reduction and Efficiency

Resource efficiency and reduction is another theme covered in this volume. The production of beer is water intensive and results in polluted wastewater. Lorr (Chapter 4) addresses water usage and treatment within craft beer production in the context of Michigan. His study provides examples of other efficiency measures, such as a recycling program of the plastic gloves used to check beer batches during the brewing process and compost agreements with local farmers to utilize the spent grain waste. Pretreatment of wastewater was also a noted initiative among one of the brewers in the study to the extent that the treatment system resulted in a

70 % reduction in wastewater production and produced even cleaner water than was originally received from the municipality.

Jones' study, in Chapter 2, portrays the understanding of sustainability by producers to involve financial management as well as the reduction of resource use, such as water, energy, and the production of waste, which is reinforced by the Brewers Association in their manuals on sustainability. Jones explains that the benefit of this type of conceptualization of sustainability is that it can be easily tracked, measured, and communicated. However, this simplified approach can penalize small producers as the cost for efficient operation equipment is high. In addition, this brings forth the notion of green marketing of craft beverages. Ironically, Jones' study identified that craft brewers were erring on the side of caution and actually downplaying their sustainability efforts.

Social Sustainability

Social capital comprises one of the three pillars of sustainability, yet appears to be far less accessible for craft beverage producers within these chapters. Local industry identity is facilitated through networks and institutions that enable insider interactions within community entrepreneurship, according to Hayward and Battle in Chapter 11. They identified the roles of intra-industry, industry-resident, and industry-outsider as essential to the development of Ashville's craft beer industry. Wright and Eaton (Chapter 5) describe how the agritourism component of cider tourism can further conserve and support the rural social fabric along with the preservation of the environment. Myles and Breen (Chapter 10) concur that artisan beverage producers help to build local and regional identities, rather than economies, particularly in the urban context. They highlight how a specific brewery is active in supporting local charities and civic organizations involving financial backing while they also host yoga classes and a community running club. Cook (Chapter 9) shows how beverage festival managers are supporting local charities and civic organizations at the local level. Yet, Jones describes less of a marketing focus on social sustainability partly because these connections are harder to quantify and do not directly impact the profitability of emerging businesses.

Slocum, in Chapter 6, examines power as it is related to social capital structures particularly involving aspects of the successful development of a beer trail in Virginia. She utilized data from craft beer producers and the tourism sector to analyze aspects of bridging social capital. Feelings of trust and safety, along with reciprocity, participation, citizen power, common values and norms, and a sense of belonging, were identified as key attributes to developing networks that support social capital building. Local small businesses support strong social capital networks with local breweries as opposed to corporate tourism conglomerates, such as hotel chains and tour companies. Destination Marketing Organizations (DMOs) are promising in their role to further development and support craft beer tourism. Slocum's work clearly outlines the need for the further development of bridging social capital, combined with government and industry leadership, particularly among diverse business structures such as alcohol producers, farms, and tourism businesses.

The Marketing of Neolocal and Sense of Place

This book advances the discussion on craft beer production as reflective of a larger societal shift toward neolocalism. Graefe, Mowen, and Graefe (Chapter 3) present a definition of neolocalism as "the feeling of belongingness to a unique local community, along with the rejection of global, national, or even regional popular culture and modernization" (pp. 30–31). They notice that the idiosyncratic characteristics of local communities are utilized and popularized in craft beverage marketing, where craft breweries are branding their products with local names, images, and history while partnering with local and environmental causes and organizations. Their timely study concludes that a strong social world exists among craft beer consumers that could potentially support messages of neolocalism. In addition, Wright and Easton, in Chapter 5, discuss how visitation to on-farm beverage production sites may be one of the few opportunities that the public has to experience agriculture. Thus, the role of craft beverage production in agricultural communities can be a positive contribution of the craft beverage tourism product. However, the idyll rural representations maintained by craft beverage producers has painted a positive image toward visitors regarding rural realities, brushing over the pervasive social, economic, and environmental challenges facing rural areas today. Hayward and Battle, in Chapter 11, attribute neolocalism and the strength of residents' attachment to place as a key component to the rise of craft breweries in North Carolina. They also note that Ashville experienced a resurgence in the preservation and redevelopment of local architecture around the time that craft beverage production began in the city.

Marketing sense of place is common practice in these research sites. Cavaliere and Albano (Chapter 7) examine the role of sense of place in craft spirit tourism marketing through the use of geographical, socio-cultural, and environmental elements of significance that are being utilized in multiple layers of craft spirit marketing and tourism. They identify that this type of craft industry marketing is contributing to bio-cultural conservation and sense of place in New Jersey. Hayward and Battle (Chapter 11) discover that local residents, tourists, and the craft beverage industry adjust their identity based on consumer feedback. Linkages to and within the community, the industry, social media, and festivals provide the key element of interaction that encourage sustainable beverage tourism development in North Carolina. In Chapter 12, O'Neill and Sozen provide deeper understandings of craft beverage development along with aspects of community and welfare considerations through an assessment of entrepreneurship. Cook (Chapter 9) explores three food festivals in Pennsylvania and further defines neolocal feelings toward craft beverages. While they are not major contributors to local economies, beer festivals provide opportunities to sample and explore craft beers and its surrounding culture. There appears to be a level of engagement between producers and consumers that reinforce the overall sense of place marketing.

Labels as forms of marketing are viewed with sometimes conflicting perspectives. For example, Wright and Easton (Chapter 5) view the idyllic and iconic labeling of cideries as examples of constructed rurality and the commodification of rural culture, whereas Cavaliere and Albano (Chapter 7) discuss the use of geographical and cultural examples on distilleries' labels as a form of local sense of place redevelopment in a peri-urban context. Myles and Breen recognize, in Chapter 10, the role of terroir in craft beer marketing and packaging through the presence of imagery, such as maps and landscapes, that support the notion that brewing is a vehicle for sustainability and neolocalism. Their work also highlights the role that craft breweries serve in contributing to sense of place thorough the repurposing of a light industrial zone close to urban centers for both production and consumption. At the same time, these previously economically marginal properties allow for the business to take advantage of lower rents while focusing on production.

Emerging Themes for Future Research

The research within this second volume situates the craft beverage industry in today's context; however, it also brings to light the unique and exciting challenges that are embedded within the craft beveragescape—challenges that may offer some possible directions for further critical research. We highlight a few of these interesting and potentially contentious themes that can guide future critical studies of the craft beverage industry.

Cooperation Versus Competition

As Hayward and Battle (Chapter 11) articulate, inter-industry communication allows for sustainable development of the craft beverage industry. Further research regarding aspects of collaboration within the industry can bring forth more detailed understandings of what cooperation means to sustainable tourism and particularly sustainable craft beverage tourism development. Connectivity (virtual via social media or directly through community and craftsperson engagement) is another way to support cooperation in the craft industry. In Chapter 3, Graefe, Mowen, and Graefe note the large number of respondents that report connecting with craft beer blogs, social media, or brewery websites. Further, this personal connection is even more evident in that over 60 % of participants said they had conversed about craft beer with a stranger they met at a brewery. Social bonding such as this is an inherent indication of neolocalism present among craft beer consumers. Slocum's research (Chapter 6) demonstrates that a high level of cooperation, in the form of social capital, occurs between craft brewers and bed-and-breakfasts. This is an interesting example of how local subject matter experts (SMEs) can form supporting networks that result in viability of both (or multiple) businesses and sectors. Moving forward, additional analysis of cooperative and co-supportive networks could prove valuable for the craft beverage tourism sector.

Additional research in cooperative efforts internally and externally to craft beverage tourism could also further support notions of anti-corporate competition and a move away from neoliberalism within this craft sector. Jones in Chapter 2 notes that the culture of craft brewing is seen by the craft beer industry as cooperative rather than competitive. This warrants future additional research as it relates to sustainability and a trend away from neoliberal economies of competition. Curtis, Bosworth, and Slocum, in Chapter 8, also deliver results that highlight the importance of linking drink-related activities with nearby recreational and cultural activities. Their work should also serve to encourage continued research into cooperation in the craft beverage sector.

In addition, cooperative approaches could further DMO research and sustainable destination management strategies. Myles and Breen (Chapter 10) note the importance of cooperation among competing breweries and the need for equity in the distribution of financial and cultural resources among craft breweries to support success. They purpose the term of coopetition (cooperative competition) in their research, explaining that breweries can serve as change agents collectively. Slocum (Chapter 6) echoes the importance of DMO involvement to facilitate communication and to create a craft beverage destination. Hayward and Battle (Chapter 11) identify intra-industry social networks, local trade institutions, tasting rooms, tours, and festivals as critical factors for identity development in craft beers. This articulation is a more nuanced form of cooperation that could serve to further applied and industry research in other geographical locations seeking to establish sustainable craft beverage tourism development.

The Political

The political aspect of alcohol consumption and production has been reiterated throughout both volumes of this series. Changes in laws, both historically and into modernity, impact consumption and production behavior in both overt and subtle ways within the US context. From a more philosophical perspective, and one that is supportive of the food tourism literature, the production and consumption of drink (and food) is an inherent political act involving ethics, economy, environment, and law. Therefore, aspects of social justice and "people/planet/profit" are critical elements that stand to be further investigated within the craft beveragescape.

It is possible that every purchase by either the producer for ingredients or end consumer is indeed a political action with political ramifications, which could be a core essence of the craft-turn. For example, Jones in Chapter 2 noted that the political impacts (local legislation, national declarations) of sustainability management from the brewers' perspective are mentioned far less frequently than the environmental measurements and impacts of their businesses. Lorr in Chapter 4 uses the political example of brewers creating a "No Fracking Way" clean water craft beer initiative to highlight environmental problems with the dangerous extraction process that threatens clean water supplies. Lorr's study also confirms a trend to stay "apolitical" in the brewers in Michigan and that this may indeed be a weak point of their sustainability initiatives regarding how they address the social justice and social equity components of sustainability.

Another example was found in Wright and Easton's study (Chapter 5) who explain that agritourism related to craft beverage tourism can potentially serve as a political impetus for rural agricultural worker to "re-emerge as authoritative spokespersons of rural life—roles currently held by outsiders, such as policy makers or actors further up the supply chain" (p. 67) as a way to engage politics in craft beverage production and consumption. Like farm tourism, craft beverage tourism (which is highly dependent on agricultural production) is inherently embedded and perhaps even embodied by political underpinnings of society.

Myles and Breen comment, in Chapter 10, that changes in legislation and local laws could both positively or negatively impact the breweries she studied. This is an example of how all states across the USA will continue to face changes in regulatory landscapes regarding craft beverage production. Future research into regulatory changes along with the potential community, business, and consumer impacts will be of interest to both academic and the applied craft beverage arena.

Authenticity of Craft Production

Understanding the genuine essences of a craftsperson or of a community allows for a truer spirit of place to emerge. Authenticity, as opposed to the idealization that often can occur in the marketing of tourism products, is a core component to sustainable tourism. Perhaps a more clear and intentional emphasis on the realities and struggles that are faced in craft beverage production and the interrelated agricultural sectors could serve to elucidate notions of future trends and obstacles faced by craft beverage destinations. Cider mills and their influence on modernity's rural cultural representation is an interesting theme that can be further developed in future craft beverage tourism research. Wright and Eaton's chapter (Chapter 5) highlights the danger in the misrepresentation of modern rural life, rural idealization and nostalgia by highlighting that tourism and agricultural interfaces could generate more accurate consumer understandings and influence their consumptive patterning to support more sustainable purchasing decisions. Myles and Breen (Chapter 10) showcase the urban reality of how brewing can serve as a vehicle of "material transformation" by taking rural agricultural inputs and transforming them into an urban product. Likewise, their work showcases how locations can be transformed from previously undesirable areas to those of vitality. Hayward and Battle in Chapter 11 explore further notions of identity and authenticity creation through feedback that occurs when the tourist and resident images agree. They report that authenticity can happen when residents and tourists observe and interact with each other in ways that allow them to compare each other's experiences. The continuum of authenticity and reproduction/idealization deserves attention in future research studies.

Bio-cultural Conservation

In Chapter 7, Cavaliere and Albano present epistemological inferences regarding the juxtapositions of localization versus corporatization and artisanal craft production versus homogenized manufacturing as related to the impacts on bio-cultural conservation in the peri-urban context of New Jersey. As craft beverage tourism research progresses, it could be important to continue to examine how local craft beverage-based enterprises, including tourism, may be contributing to bio-cultural conservation, particularly as compared to corporate conglomerate beverage producers. For example, Jones in Chapter 2 makes the astute point that the trend toward CS can be understood as the "McDonaldization" of CSR because aspects like social capital and human rights, that are far more challenging to measure, have become marginalized while the pursuit of the more easily quantifiable aspects of resource management are emphasized.

Craft producers are uniquely positioned to grant value to these bio-cultural connections to conservation—that society and community are integral components to just, fair, and sustainable business development. Unlike corporate conglomerates that are driven by efficiency and cost-saving strategies, small entrepreneurial craft-based beverage producers can organically weave care for people into the planet and profit discourse. Jones suggests that notions of commitment, creativity, integrity, diversity, and other similarly qualitative components be celebrated as unique and marketable aspects of sustainability to promote craft beverage production. These aspects warrant additional research, particularly within the emerging field of craft beverage tourism.

Understanding the Craft Beverage Tourist

Graefe, Mowen, and Graefe (Chapter 3) suggest that increasing the importance of localism and making stronger connections between craft beer and environmentalism issues may result in expanded environmentalism among consumers. Using the Lifestyles of Health and Sustainability (LOHAS) framework, they emphasize a market segment that focuses on environmental, social, and personal health attributes as they relate to conscious capitalism and post-modern ethics. Curtis, Bosworth, and Slocum (Chapter 8) highlight already existing connections between outdoor enjoyment and craft beverage consumption. Therefore, additional research on the craft beverage consumer in relation to their affinity for healthy lifestyles can uncover other psychographics, core values and consumption habits.

As noted by Jones, in Chapter 2, the mainstream beer industry has historically been related to constructions of masculinity. The marketing and sustainability ramifications of gender in craft beverage production and consumption can benefit from deeper investigations. Wright and Eaton (Chapter 5) recognized that the name of farms using family names "serves to masculinize the cider mill, elevating male heads of households over women entrepreneurs" (p. 72), which undervalues women's identity and contributions to community sustainability and indeed entrepreneurship in general. Cook (Chapter 9) also recognizes the role that gender plays in attendance and participation at craft beer festivals as an often overlooked and rarely addressed aspect of research. His chapter specifically analyzes gender as an element of consideration regarding craft beer festivals in Pennsylvania. He finds that large numbers of women attend the festivals and are extremely knowledgeable craft beer tourists. Future research could further his work by incorporating more detailed gender considerations into research design and analysis.

Additionally, Graefe, Mowen, and Graefe note in Chapter 3 that there has been a growth in consumption of craft beer among Hispanics and those with lower incomes. The craft beverage movement has been noted as primarily a White one—from both the supply and demand side (Withers, 2017). However, without adequate consumer research exploration, our understanding is sorely limited.

Last Call

As our work on a double-volume set of craft beverage research comes to a close, we wish to encourage academics and industry professionals to continue the scholarship that has only been broached on craft beverage tourism. Everett and Slocum (2013) recognize that food and beverage tourism is not a unified phenomenon, but that each branch offers subtle insights into the overall phenomenon of cultural exploration through taste. We hope these books offer a foundation on which to guide future academic endeavors, while realizing that more questions surfaced than have been answered.

While this work has centered around craft beverage production and tourism in the USA, primarily because of the rapid changes in alcohol regulations and partly because of the increasing interest in neolocal production that is sweeping the country, we await additional insights from outside the USA as a means to inform best practice and further appreciate this interesting trend. Craft beverage is a fun and exciting topic, tantalizing not only to our taste buds but also our scholarly curiosity and intellect. It is featured prominently in the push toward localism and regional development, as well as offering opportunities to support the craft movement and traditional industries. No matter what your native language, I am sure you recognize these calls to action:

Cheers
Prost
Salute
Santé
干杯
صحتك في
На здоровье
Skål
건배
And our personal favorite—Bottoms Up!

References

Everett, S., & Slocum, S. L. (2013). Food and tourism, an effective partnership? A UK based review. *Journal of Sustainable Tourism, 21*(7), 789–809.

Hoalst-Pullen, N., Patterson, M. W., Mattord, R. A., & Vest, M. D. (2014). Sustainability trends in the regional craft beer industry. In M. Patterson & N. Hoalst-Pullen (Eds.), *Geography of beer* (pp. 109–118). New York, NY: Springer.

Withers, E. (2017). The Impact and Implication of Craft Beer Research: An interdisciplinary literature review. In C. Kline, S. L. Slocum, & C. T. Cavaliere (Eds.), *Craft beverages and tourism, Volume 1: The rise of breweries and distilleries in the United States* (pp. 11–24). New York: Palgrave Macmillan.

Index

A
agritourism, 65, 73, 220
Asheville, N.C., 44, 171
authentic experience, 119, 172, 187, 189

B
Beer City, USA!, 50, 171, 181–3, 186, 188
beer festivals, 153–5, 171, 184–7, 218
Bell, 56, 58, 59
bio-cultural conservation, 7, 102, 114, 214, 218, 222
brewery tours, 16, 85, 94, 186
Brewery Vivant, 12–14, 18, 50, 52–5, 57, 60–2
business environments, 4, 29, 63, 199, 208

C
carbon footprint, 60
case study, 50, 88, 90, 144
charitable cause, 155
cider mills, 4, 65, 73, 221, 223
collaboration, 42, 58, 84, 85, 96, 164, 165, 219
community building, 162
consumption, 2, 5–7, 14, 15, 20, 27, 28, 42, 50, 67, 75, 76, 83, 85, 110, 135, 160, 162, 166, 167, 218, 220, 221, 223
corporate social responsibility (CSR), 2, 3, 10, 11, 14, 17, 22, 23, 51, 62, 214, 222
corporate sustainability (CS), 3, 10, 11, 14, 22, 23, 214, 222
craft beer, 215–20, 222, 223
craft beverage, 1–3, 5–8, 33, 75, 77, 102, 112, 114, 168, 213–24
craft brewery, 13, 27, 31, 35, 53, 107, 141, 142, 154, 174–6, 184
craft brewing, 3, 9–11, 14–16, 19, 21, 22, 50, 51, 54, 62, 141, 172, 175, 195, 219
craft distilleries, 101
CSR. *See* corporate social responsibility (CSR)

D

demographics, 120–3, 127, 128, 135, 144, 147, 199, 202
development, 2, 6–8, 66, 68, 85–7, 90, 94, 95, 98, 101–3, 112, 120, 137, 161, 166–7, 172, 174, 176, 177, 181, 186, 187, 189, 195, 197, 202, 208, 209, 214, 216–20, 222, 224
drink festivals, 6, 33, 34, 36, 42, 120, 122, 126, 135, 142–51, 153–5, 171, 183, 185–7, 218, 223
drink tourists, 5, 120–2, 127–9, 132, 135, 136, 214, 215

E

economic impacts, 83, 120, 168
effluent, 56, 58, 60
Enbridge, 49, 52
entrepreneurs, 2, 5, 67, 72, 114, 165, 166, 172, 174, 176–8, 184, 195, 197–200, 205, 208–10, 223
environmental, 10–20, 22–4, 27, 35–9, 41, 49–53, 55–7, 62, 63, 79, 105, 106, 161, 168, 196, 198, 215, 217, 220, 222
environmental attitudes/values, 4, 30, 33–5, 39, 40, 43–5
environmental behaviors, 31, 44, 45

F

food tourism, 2, 3, 7, 84, 85, 87, 91, 93, 95, 119, 121, 137, 220
founders, 50, 52–8, 62

G

gentrification, 166, 167
Great Lakes, 16, 17, 49, 51, 69

green marketing, 21, 54, 216
greenwashing, 3, 10, 12, 16, 21

H

Human Agency, 209

I

Intermountain West, 120, 122, 123, 135, 136, 214

L

Lifestyles of Health and Sustainability (LOHAS), 32, 33, 222
local, 2–7, 12, 19, 20, 22, 24, 28–31, 33, 35, 36, 38, 40, 42–4, 52, 53, 55–60, 63, 66, 67, 70, 74, 75, 80, 84, 85, 91–8, 104, 107, 111, 112, 114, 119, 126, 127, 130, 132, 136, 137, 142, 144, 145, 147, 148, 152, 154, 155, 160–2, 164, 166–8, 171–9, 181–3, 185–90, 196, 199, 202, 208, 215–22
local food, 2, 5, 42, 68, 126–8, 130, 132, 136, 175, 215
local identity, 6, 31, 172–5, 178, 183, 187, 190
local image, 187, 188
local production, 5, 103, 114
Loudoun County, 83, 84, 87, 88, 91, 92, 96–8

M

McDonaldization, 23, 222
Michigan Brewers Guild, 53, 61, 63
microbrewery, 30, 31, 143, 163, 176
motivation, 3, 5, 7, 8, 15, 16, 23, 29, 59, 92, 122, 135, 184, 195

N

neolocalism, 4, 27, 29–31, 33, 35–44, 67, 160, 175, 176, 214, 217–19
networking, 90, 91, 96, 98
New Belgium, 12–14, 19, 172, 188, 189
New Jersey, 5, 101–5, 107–12, 114, 218, 222

P

packaging, 14, 76, 160, 174, 218
Pennsylvania, 4, 6, 31, 33, 34, 42, 141, 142, 144–6, 153, 215, 218, 223
production, 1–7, 27, 55, 59, 67, 69, 70, 74–80, 83, 84, 91, 95, 101–3, 109, 110, 112, 114, 136, 141, 142, 160, 162, 166, 168, 175, 184, 214–18, 220–4
prohibition, 7, 101, 102, 108, 109, 142, 143
psychographics, 120–2, 127, 128, 223
Pure Michigan, 49

Q

qualitative, 23, 24, 84, 87, 104, 120, 222

R

(re)development, 101, 102, 161, 166
representation, 4, 65–9, 72–4, 79, 80
residents, 6, 7, 61, 87, 122, 148, 164, 166, 167, 172, 173, 176, 177, 181–3, 185–90, 218, 221
rural development, 66, 71
rural idyll, 65, 66, 77–9
rurality, 65, 69, 71, 73, 75, 218

S

sense of place, 1–3, 5, 7, 31, 67, 101–6, 113, 114, 176, 186, 214, 217
short, 2, 15, 68, 69, 144–6, 154, 168, 174
Sierra Nevada, 12–14, 59, 172, 188, 189
small and medium enterprise (SME), 2, 7, 102, 103, 106
social capital, 2, 4, 5, 23, 83–7, 90–2, 96–8, 214, 216, 219, 222
social interaction, 36, 43, 69, 172, 173
socio cultural, 2, 106, 199
spirits, 1, 2, 5, 30, 101–3, 107–9, 111, 112, 126, 141, 196, 214, 217, 218, 221
sustainability, 2–5, 9, 28, 30–5, 38, 40, 49, 51, 72, 85, 101, 160, 168, 213–16, 218–20, 222, 223

T

tasting rooms, 6, 70, 76, 78, 87, 179, 183, 184, 189, 220
tourism, 1, 4–8, 28, 29, 34, 42, 44, 49, 66–70, 72, 75, 80, 83, 104–5, 112, 114, 119, 155, 160, 161, 165–8, 172, 173, 175, 176, 181, 183–5, 187–90, 214, 216–24
tourism industry, 4, 84, 85, 87, 88, 95, 104, 160, 176, 183
tourism marketing, 5, 28, 42, 184, 217
tourism trails, 159, 161, 162

tourist profile, 119
transportation, 14, 85, 101, 188
triple bottom line, 10, 51, 62

U
Utah, 5, 122, 125–30, 132, 135, 137, 142

V
Virginia, 4, 83, 85–8, 92, 216

W
water pollution, 50, 52, 59
website content analysis, 102, 105, 106

Printed by Printforce, the Netherlands